A Philosophy of Cover Songs

P.D. Magnus

OpenBook Publishers

https://www.openbookpublishers.com

©2022 P.D. Magnus

ISBN Paperback: 9781800644229

ISBN Hardback: 9781800644236

ISBN Digital (PDF): 9781800644243

ISBN Digital ebook (epub): 9781800644250

ISBN Digital ebook (azw3): 9781800644267

ISBN Digital (XML): 9781800644274

ISBN Digital (HTML): 9781800646773

DOI: 10.11647/OBP.0293

The cover image is a collage of AI-manipulations on the photograph 'Yes Music in the Amphitheater, 1970' by Ed Uthman, CC-BY-SA 2.0

Cover design by Anna Gatti

Contents

List of Figures

To Cristyn, Christy, and Ron.
Their influence runs through the whole book.
Even though the words are mine, we found many of the ideas together.

Acknowledgments

In the late 20th century: With the resource of Napster, my brother Warren put together two dozen versions of the song 'Stardust.' Inspired by this, I assembled several similar compilations— 29 versions of 'Don't Get Around Much Anymore', 12 versions of 'Solitude', and 8 versions of the theme from *Shaft*. I had great fun listening to different covers, exploring the potential of each song.

About a decade ago: My friend Christy Mag Uidhir reached out to me, said that he had some ideas about covers, and suggested that we should coauthor something. We worked at it but disagreed on some fundamental issues. My wife Cristyn and I had long conversations in which it became clear that neither Christy nor I had staked out the right position. Distinctions needed to be made. That collaboration led to a paper by the three of us (Magnus et al. 2013). When responding to that paper, Andrew Kania refers to the three of us as 'the Mags'— he says, 'to save some space!' (2020: 238)

More recently: Christy suggested that it was time to say more about covers. Sure, I replied, but what should we say? He didn't know. I brainstormed lots of possibilities. The three of us soon had a draft of a new paper. I was having weekly lunch with Ron McClamrock, who used to play in cover bands. He brought a different perspective to the project and ultimately joined us as a coauthor (Magnus et al. 2022). It remains to be seen how this collaboration will be abbreviated— perhaps '4M' or 'the Mags feat. Ron McM.'

In the course of this collaboration, I came up with more ideas than would fit and wrote things which were cut for space. I had sabbatical coming up, giving me time to think about the bigger picture. The broader ideas and stray drafts formed the basis of this book.

Thanks to the Department of Philosophy at the University at Albany for making this book possible— for the semester of sabbatical in Fall 2021 which gave me a chance to focus on the project, but also for the willingness to tolerate my dilettantism.

The account I give here is built on the one I developed with my coauthors, and it would not have been possible without them. For feedback on drafts of the book, thanks to Christy Mag Uidhir, Ron McClamrock, Evan Malone, and Warren Magnus. Thanks also to Oz McClamrock, whose secondhand input about musical practice was useful at many stages of the project.

P.D. Magnus
Albany, 2022

'Now you're telling me you're not nostalgic.
Then give me another word for it.
You who are so good with words
And at keeping things vague.'
— Joan Baez, 'Diamonds and Rust' (1975)

'Now you're telling me you're not nostalgic.
Then give me another word for it.
You who's so good with words
And at keeping things vague.'
— Judas Priest, covering Joan Baez (1977)

'Now you're telling me you're not nostalgic.
Well, then give me another word for it.
You were so good with words
And at keeping things vague.'
— Great White, as a tribute to Judas Priest (2008)

Introduction

Cover songs are a familiar feature of contemporary popular music. Musicians describe their own performances as *covers*, and audiences use the category to organize their listening and appreciation. However, philosophers have not had much to say about them.

A common philosophical approach is to consider historical positions—for example, asking what Plato or Kant said on a topic. That makes no headway here, because Plato and Kant had nothing to say about cover songs. How could they? A cover is a version of a song that was first recorded by someone else, so covers require the technology to record and play back music. If Plato or Kant wanted to listen to a song again, they had to find performers to play it again.

Nevertheless, philosophy provides a valuable toolbox for thinking about covers, and the philosophy of cover songs illustrates some general points about philosophical method. Why is it that people have been announcing the death of covers for as long as there have been covers, while musicians keep making them? To answer that, we need to introduce distinctions. There are different kinds of covers.

As much as we need distinctions, however, we also need to recognize that honing our categories to diamond precision can be pedantic and doctrinaire. There are some distinctions which are not worth making.

A philosophical account of cover songs would be perverse if it were just an ethereal abstraction, so I discuss lots of different examples in this book. You may already be familiar with some of them. Others may be new to you. When it matters what a particular record sounds like, there is no reason to take my word for it. Most of the recordings that I discuss are readily available on the internet. You should listen and decide for yourself.

Please keep two things in mind about the examples.

First, you may find that you have different opinions about some of them than I do. Where the disagreement is incidental to the broader philosophical

 https://doi.org/10.11647/OBP.0293.09

point, I ask you to substitute an example which you find more agreeable.

Second, despite all of the examples I address, there are many more that I have had to leave out. If you find yourself thinking about examples that I do not explicitly discuss, then I invite you to apply the distinctions and moves I make in the book. Part of the fun of thinking about cover songs is that there are interesting examples all over the place.

This book is divided into three parts, each containing two chapters. Although issues recur at different places, I have tried to make the chapters stand coherently on their own.

The first part is about *how to think about covers*. Chapter 1 reviews the history of covers and topples some possible definitions of 'cover.' Even though there is no clear definition, we can get by without one. I take cover songs to be the ones that are typically called that, and that is enough to get going. Chapter 2 introduces several distinctions which can help us understand covers better: First, between songs, performances, and tracks. Second, between mimic covers and rendition covers.

The second part is about *appreciating covers*. Chapter 3 uses the difference between mimic and rendition covers as the key to thinking about how we evaluate and appreciate them. Chapter 4 discusses covers which have an especially strong connection to the original— the original is either alluded to by the cover or changes how we hear the cover. Evaluating and appreciating these covers turns out to be tricky.

The third part is about *the metaphysics of covers and songs*. Chapter 5 poses some puzzles about the metaphysics of cover versions. Although a cover is typically a version of the same song as the original, there are some interesting and striking counterexamples. Chapter 6 develops an analogy between songs and biological species. Like species, songs are historical individuals. This shows how to resolve the puzzle cases from the previous chapter and also helps in thinking about oddities like mashups, parodies, and instrumental covers.

1. What is a Cover?

Consider this recent Twitter thread:

> 8:00 PM Jun 3, 2021
>
> **Panda Lakshmi:** Once I was singing Istanbul (not Constantinople) in my house and my mom starting singing along. Turns out the TMBG version is a cover!
>
> **Ellen Fuoto:** So my 70 something year old brain is starting to wake up. You mean They Might Be Giants made a cover of the Four Lads hit from back when I was 6?
>
> **Uglysquirl:** You two have blown my mind.. I'm a big fan of cover songs and TMBG and I never knew this was a cover.
>
> **gargoyle:** I... With the.. But... I was today years old when I learned this. I'm going to need some alone time to deal with this crisis. ☺

The song that they are talking about ('Istanbul (Not Constantinople)') was a hit in 1953 for the quartet the Four Lads, and it 'found its way into our cultural lexicography as one of those songs that you knew you knew, but didn't know where you knew it from' (Treble 2018). The duo They Might Be Giants (TMBG) recorded a faster, livelier version for their breakthrough 1990 album *Flood*. The younger participants in this thread are surprised to learn this, because they just know it as a TMBG song. At the same time, the one older participant is surprised to learn about the TMBG version. Minds are blown. Even though the winking emoji in the last post makes clear that nobody's life is deeply changed by this discovery, all the participants find it significant that the TMBG version is a cover.

Music audiences, which include you and me, use the concept of *cover* to understand certain songs, performances, and recordings. We take the difference between original and cover to be significant. But what is the difference? What does 'cover' mean?

https://doi.org/10.11647/OBP.0293.01

The dictionary definition

A 'cover' is typically defined as a recording of a song that was first recorded by someone else. Something like this is given in many dictionaries and by some scholars. For example: Albin Zak provides a glossary entry defining a 'cover version' as 'A recording of a song that has been recorded previously by another artist' (2001: 222). Don Cusic writes, 'The definition of a "cover" song is one that has been recorded before' (2005: 174).

If it were that simple, this could be a short book. Inevitably, complications arise. Let's look at five of them.

Five problems

1.   Consider the song 'Let It Be', written by John Lennon and Paul McCartney. Their band, the Beatles, had a hit with it when they released their version in 1970. However, the first released recording of the song— by a few months— was by Aretha Franklin. A website which generates its descriptions automatically labels the Beatles' version as a cover of Franklin's, and that is just what the usual definition would suggest. However, this seems absurd. If either version is a cover, then it is Franklin's. Lennon and McCartney were members of the Beatles who wrote the song with the intention of recording it, even though McCartney sent a demo to Franklin in hopes that she might record a version. It just happened that her version was released earlier.

One might think that the prior existence of the demo makes Franklin's version a cover, but many recordings— most in recent decades— exist as demos before there is a published version. To take just one example, consider Patsy Cline's 1961 hit 'Crazy.' The song was written by Willie Nelson, who was trying to get a singer to record and release it. He cut a demo record of 'Crazy' and played it in a bar in Nashville for Patsy Cline's husband, who insisted he play it for Cline. She loved it and recorded her version the next week. Although Nelson had recorded a demo, almost nobody calls Cline's version a cover. It does not show up on internet lists of *best cover songs* or *songs you didn't know were covers*. Artists on YouTube typically list their versions of 'Crazy' as covers of Patsy Cline. So the existence of a demo does not seem to make Cline's version a cover.

However, consider 'Girls Just Want to Have Fun', a hit for Cyndi Lauper in 1983. It was written by Robert Hazard, and he recorded a demo version in 1979. Surprisingly, Lauper's version appears on many of those internet lists. This may partly be confusion because Hazard's demo was later published (to piggyback on the success of Lauper's version) but often these lists acknowledge that Hazard's version was a demo. One comments, 'Hazard's recording never got past the demo stage, so I'll choose to consider Lauper's version "technically a cover but sort of not really"' (Proximo 2017). When Lauper's album was selected for the National Recording Registry, a webpage at the Library of Congress included the comment, 'Lauper's take on Robert Hazard's "Girls Just Want to Have Fun" wasn't a mere cover, it was a transformation of the song into a joyous feminist anthem' (NRPB 2018). Something which is not a mere cover is more than a cover, rather than not being a cover at all.

Another example is the Crickets' 1957 hit 'Oh Boy', which is often described as a cover of Sonny West's version (Londergan 2018). West, who cowrote the song, had recorded a demo of it under the title 'All of My Love.'

Contrary to the simple definition, the existence of a demo version does not automatically make a version a cover— but maybe it does sometimes. Call this *the problem of demo versions*.

2. If we accept Cline's version of 'Crazy' as the original, then later recordings should count as covers. However, when Willie Nelson recorded it for his debut album the following year, it was not obviously a cover. Here common usage is unclear. Some people count Nelson's version as a cover (due to Cline's original) but others do not (due to Nelson having written the song). It is a vexed question.

It is also unclear how to think of cowritten songs. Consider two cases: First, the song 'China Girl' was cowritten by Iggy Pop and David Bowie, and Bowie played on and produced Pop's 1977 recording. Bowie recorded his own version in 1983 without Pop. Second, the song 'Layla' was cowritten by Eric Clapton and Jim Gordon. They recorded it with their band (Derek and the Dominos) in 1971. For MTV Unplugged in 1992, Clapton recorded an acoustic version which won the Grammy Award for Best Rock Song. The usual definition would call Bowie's 'China Girl' and Clapton's 'Layla' covers, and some people would agree (Leszczak 2014, Popdose 2011). Bob Leszczak, for example, describes the MTV Unplugged performance as Clapton having 'covered his own song' (2014: 120). Other people are inclined to say that these are not covers.

So there are vexing questions about whether and how a person who wrote or cowrote a song can cover earlier recordings of it. Call this *the problem of songwriters*.

3.   The typical definition only applies to new recordings. Yet a cover band is a musical group that just performs covers, and most cover bands perform live rather than recording. This shows that the word 'cover' is readily applied to live versions as well. There is an asymmetry, however, because something is not a cover if it is a recorded version of a song that has previously been performed live. Even though a cover may be a live version, the earlier original must be a recording. Call this *the problem of live versions*.

4.   A cover of a song need not include any singing, and instrumental versions are regularly labelled as instrumental covers. Nevertheless, one would not call Miles Davis and Cannonball Adderly's 1958 version of 'Autumn Leaves' a cover. There were earlier released recordings by Yves Montand in 1946 (as 'Les feuilles mortes'), by Dizzy Gillespie and Johnny Richards in 1950 (as the instrumental 'Lullaby of the Leaves'), and by others. The tune had become a jazz standard. When it is played today, one might compare the new performance to the famous Davis/Adderly version, but one would not call the new performance a cover.

'Autumn Leaves' is not extraordinary in this regard. We treat jazz recordings differently than we treat rock recordings. This has prompted writers like Deena Weinstein (1998) and Gabriel Solis (2010) to argue that covers only exist in rock music. Weinstein writes, 'Cover songs, in the fullest sense of the term, are peculiar to rock music, both for technological and ideological reasons' (1998: 138). However, this requires an expansive conception of what counts as rock. There are covers in pop music and contemporary country as well. Moreover, there are numerous points of interaction between jazz and rock (especially rock in this expansive sense). It is unclear how to draw the boundaries around the regions of musical or cultural space where covers are possible. Call this *the problem of genre*.

5.   Contrast two cases: First, Kid Cudi's 2008 '50 Ways to Make a Record' follows the same melody and musical structure as Paul Simon's 1975 '50 Ways to Leave Your Lover' but replaces Simon's lyrics about love lost with ones about making music. Cudi's track is often described as a cover of Simon's. Second, Weird Al Yankovic's 1981 'Another One Rides the Bus' follows the melody and musical structure of Queen's 'Another One Bites the Dust' but replaces lyrics about being indomitable with ones about public transit. 'Another One Rides the Bus' is usually described as a parody of 'Another One Bites the Dust' and is *not* counted as a cover.

The percentage of words shared between the original and the parody does not seem to matter. There is parallel structure in the title and lyrics between Cudi and Simon but also between Yankovic and Queen. Perhaps the only thing which stops 'Another One Rides the Bus' from being a cover is that it is a parody, which in turn is because it is funny. And '50 Ways to Make a Record' counts as a cover because it is not a parody, which in turn is because it is not funny. John P. Thomerson, who denies that parodies have to be humorous, seems to count all covers as parodies; he refers to typical cover band performances as 'reverential parodies of classic rock and country hits' (2017: 1). A definition of 'cover' should be able to make sense of this. Call this *the problem of parodies*.

Looking for the real definition

These problems are reasons to be unhappy with the usual definition, and we can use them as a toolbox to dismantle other definitions. For example, Andrew Kania defines a cover in this way: 'A cover version is a track (successfully) intended to manifest the same song as some other track' (2006: 412). This is vulnerable to all of the problems discussed above.

One might start tinkering with these definitions, adding clauses to resolve each of the problems. Yet that is not the only possible response.

An alternative approach supposes that the meaning of the term is determined by how it was introduced. The word 'cover' refers to a particular type of thing. So it has a real definition, the true nature of those things, regardless of what ordinary people or scholars might say when asked to define the term.

This approach was originally applied to proper names and to natural kind terms like 'gold.' The idea is that the word 'gold' was introduced to describe samples of gold, and it meant *that kind of stuff*. For centuries, people did not know what gold really was. They could not have given a true and informative definition. Only later did chemists develop atomic theory and physicists learn the structure of atoms, allowing us to characterize gold as a chemical element in terms of the number of protons in each of its atoms.

Nevertheless, that is what 'gold' meant all along (on this account). (My gloss of the view here is rather breezy. Key texts are by Saul Kripke (1972) and Hilary Putnam (1975), and decades of literature have followed.)

Although the category of cover versions does not look like a natural kind, it has also been suggested that this approach to meaning applies to artifacts (Putman 1982). So maybe 'cover song' picks out *that kind of recording* or *that kind of version*. Unlike 'gold', which entered Old English from even older sources, the word 'cover' in the sense that interests us arose only in the late 1940s. So let's turn away from puzzle cases and consider some history.

The history of covers

The term 'cover' first found widespread usage in the 1950s, corresponding to a shift in the record business.

Here is the simplified version: Before the 1950s, songs which everyone played became *standards*. This is natural when the paradigm case of music was live performance, both because performance is ephemeral and because it is done by whatever musicians someone has in front of them just at that time. Radio, initially dominated by live performance, did not immediately change this paradigm. After the 1950s, new versions of songs are often considered in relation to earlier recordings of that same song which are taken as canonical or original. The new versions are *covers*.

Early days

Initially, customers tended to seek out a particular song rather than a particular recording of that song by a particular artist. By covering a song, a record company could steal sales which would have gone to a competitor. As John Covach and Andrew Flory write, 'When the original version appeared on a small independent label, a larger independent label (or a major label) could record a cover and distribute its records faster and more widely....' They add that 'to some extent, this explains the greater success of these versions and why we call them "covers"' (2018: 87). Ray Padgett writes that covers in the 1950s were 'copycat recordings done quickly' and suggests two reasons these might have come to be called 'covers': First, a publisher might be '"covering its bets" by releasing its own recording of a popular song.' Second, it was aimed to '"cover up" another version of the same song on a store's shelves' (2017: 4).

That is only part of the story. In a 1949 *Billboard* magazine article on small record labels, Bill Simon writes:

> The original disking of *Why Don't You Haul Off and Love Me?*, cut for King [a small record label] by Wayne Raney, has hit 250,000, and versions are now available on all major labels. None of these, however, has approached Raney's mark. Another King disk, *Blues Stay 'Way From Me?*, by the Delmore Brothers, is close to 125,000 in six weeks, and other companies have just begun to cover the tune. (1949: 18)

Here 'cover' has the sense of *coverage*. Just as a band might try to learn the popular songs that an audience member might request, a record company wanted to be able to have a version for sale. This reflects how songs work. A song can be performed by different artists. It is not matched one-to-one to the person who wrote it or the singer who made it famous.

From a commercial standpoint, there is no reason to make something original. It is easiest just to copy the interpretation and arrangement of a hit record, and the success of the hit suggests that it might be more commercially successful than trying something new. So there was a shift from making sure a label's library covered the repertoire to cutting records that just copied successful ones. In 1954, the chain store Woolworth's launched its own record label in the UK, Embassy Records. Their entire line was cheaply recorded knock-offs (Inglis 2005, Woolworths 2017).

Some in the industry commented on the contrast between earlier covers (new versions of a song recorded for coverage) and these new copy recordings. A 1955 *Billboard* article laments 'the duplication (rather than the covering) of successful disks' (1955a). An article a few months later describes a New York radio station that 'will henceforth refuse to play "copy" records.' The article explains that this policy 'draws a clear distinction between "cover" records and "copy" records— defining the latter as those disks which copy— note for note— the arrangement and stylistic phrasing of the singer' (1955b). Nevertheless, the word 'cover' came to apply to both sorts of records— both recordings of the same song that used a different interpretation or arrangement and also those that copied the interpretation and arrangement of the original recording.

A further feature of music in this period was the centrality of rankings in trade magazines as a measure of commercial success. At least in the United

Figure 1: In the mid-1950s, Woolworth's Embassy Records built a state-of-the-art studio in London for recording cheap covers.
Image courtesy of the Woolworths Museum.

States, this introduced complexities of race and class. The *Billboard* magazine rhythm and blues (R&B) chart had, prior to 1949, gone under a succession of other titles: 'Harlem hit parade', 'race', and 'sepia.' As the earlier names make clear, the chart was not meant to capture a particular style of music but instead a particular audience demographic— black people. The country and western chart, previously 'hillbilly' and 'folk', was also organized around a particular audience. As Covach and Flory note, 'Rhythm and blues… charts followed music that was directed to black urban audiences, and country and western… charts kept track of music directed at low-income whites' (2018: 85). This left the pop charts, although nominally just tracking popular music, focused predominantly on the white, middle-class market.

Covach and Flory put the point in terms of the music's target audience, but a song could have success beyond just its target. The charts were constructed based on reports from radio stations and juke boxes (of what they were playing) and from record shops (of what they were selling). As a result, a song by a black artist could make it onto the pop charts if it had plays and sales in places to put it there. Similarly for country musicians. A song that made it onto multiple charts was called a *crossover*, and crossing over meant a distinct kind of commercial success.

Although some crossover hits were a single record appearing on multiple charts, others were the same song but recorded by different artists. Given the racial division of the charts, there are striking examples of white artists having pop hits with songs that had been R&B hits when recorded by black artists. The most famous example of this is probably Pat Boone's 1956 pop version of Little Richard's 'Tutti Frutti.' In that case, the cover by the white artist did not completely eclipse the original. Although Boone's cover reached #12 on the *Billboard* pop chart, Little Richard's reached #17 on the pop chart and #2 on the R&B chart. Regardless, this is just one instance of a broader pattern in which, as Denise Oliver Velez puts it, 'Black music… was "borrowed," "lifted," "copied," and made money for white artists, often garnering both commercial success and awards… while leaving the Black originators with far less, or nothing' (2021). Singer-songwriter Don McLean describes it this way:

> [I]f a black act had a hot record the white kids would find out
> and want to hear it on 'their' radio station. This would prompt
> the record company to bring a white act into the recording studio

and cut an exact, but white, version of the song to give to the white radio stations to play and thus keep the black act where it belonged, on black radio. A 'cover' version of a song is a racist tool. (2004)

The word 'cover' suggests itself here perhaps as a contraction of 'crossover.' McLean leverages this as a definition, to argue that Madonna's 1999 version of his 1971 song 'American Pie' should not be called a cover. Yet, common usage treats Madonna's version as a cover. Although covers were sometimes used as racist tools, racism is not intrinsic to the concept of a cover as such. As Michael Coyle puts it, crossover covering of R&B hits by white artists 'exploited racist inequality but did not arise because of it' (2002: 144).

The word cover originally had a sense of *coverage* which was not in itself tied to race, and covers in that sense continued. Even when a cover eclipsed the original, it was not always about race. For example Sonny West cowrote and recorded 'All My Love (Oh, Boy)' (1957) and 'Rave On' (1958), but both were covered by Buddy Holly and the Crickets. Borrowed, lifted, and copied, but by white musicians from a white musician.

In the earlier, song-focussed market, songwriters and publishers would make money from sheet music as well as recordings. In the 1950s, the situation was changing. The only way for a country and western song to sell successfully as sheet music was if it crossed over to the pop charts (Gabler 1955). And soon enough sheet music would not be a central concern at all, as the primary product became the recording itself. Because of the changing marketplace, covers were a way for a song to get exposure to a broader audience. This was good for songwriters (who got a royalty from every sale, regardless of whose version was selling) but bad for performers (who profited only from sales of their records).

Coyle argues that this history fails to capture what covers really are. He writes 'that no one in 1954 would have used the word "cover" to mean what we mean by it today.' The sense of the word 'cover' that I've discussed so far in this section is what Coyle prefers to call *hijacking a hit* or just *hijacking*. Although hijacking was called covering in the 1950s, Coyle maintains that the word means something different now. He writes, 'The notion of covering a song has changed radically in meaning because… the relation of writers to performers to audiences… has changed radically' (2002: 136).

Coyle maintains that the contemporary sense of 'cover' began in the late 1950s and that, 'in our modern sense of the term, Elvis Presley was the first cover artist' (2002: 153). Elvis neither wrote his own songs nor recorded ones that were current hits. Instead, he recorded songs that had faded from memory. Coyle writes, 'In recovering nearly forgotten recordings by black artists Presley was doing much more than reviving potentially money-making properties; he was using recordings by black artists to perform for himself and for America a new identity' (2002: 153). Writing about subsequent developments in the late 1960s, Coyle writes that 'while the black audiences for 50s-style R&B had long since moved on to other styles, there was an audience of "serious" white fans' eager to embrace a blues revival (2002: 152). Elvis also recorded covers of country songs, but the R&B songs did more to define his image.

The covers that Coyle highlights exploited race in a different way than McLean describes. Whereas Pat Boone recorded songs written by black musicians without any suggestion of their origins, Elvis and later artists positioned themselves explicitly as white musicians performing black music. So, Coyle claims, covers were 'a way for performers to signify difference' and to 'project their identity' (2002: 134). This identity was bound up with issues of race, because 'white groups were striving to sound black' by harking 'back to material that black audiences had already largely abandoned' (2002: 143).

So Coyle advances two theses. The first is that the early-50s sense of 'cover' went away. The second is that it was replaced by 'cover' in the sense of a recording that establishes the recording artist's identity by signifying the original version in a way that exploits the dynamics of race. Although he is pointing to important historical developments, neither thesis is true. I will explain why in the next two sections.

Hijacking continues

Coyle is right that there were changes in the music industry in the late 1950s which made covering (in the sense of what he calls hijacking) less prevalent. However, it did not go away. The Scottish jazz musician Sandy Brown still defined 'cover' in those terms in 1968; he writes, 'The jackal thinking behind cover versions, which are near copies of original recordings, is predicated on the belief that so much money is showered in the general direction of hit records that any performance of the song will collect if sufficiently adjacent' (1968: 622). Adapting Brown's language, we might call these *jackal covers*. They continue to be at least part of what contemporary audiences think of as covering.

If we look at the music industry press, there have been declarations that covering in that sense was on the way out for almost as long as there have been covers. Considering the success of Decca records, Milt Gabler writes in 1955, 'The day of the fast, haphazard "cover" record is gone. This does nothing but lose money for the company, the artist and the publisher. To-day more money is put into advertising and exploitation than at any other period in the history of the business. Records must be good to pay off.' He adds, 'The best chance a new artist has is with new material or an outstanding arrangement of a great standard!' (1955) An article in *Cashbox* magazine a couple of years later discusses changes underway in the music business: 'Record fans in the current market know the records of all the fields and very often even if there are cover records, they want the original one' (1957). In 1970, the head of a record label is reported in *Billboard* to have said that covering was 'a costly affair' because 'a company that comes out with a "cover record" has to put an extra effort to beat the original and this means a heftier outlay in promotion and advertising expenditure' (1970).

Take one vivid example: A&M records released 'Fugitive', a guitar instrumental by Jan Davis. Dolton records released a cover by the Ventures, taking out a full-page ad in the April 11, 1964 issue of *Cashbox* magazine which announced that the Dolton disk was 'Running headlong for the charts!' Since neither version of 'Fugitive' made it into the *Billboard* Hot 100, maybe it just shows that you cannot hijack a hit if your target does not end up being a hit— but there is more. A&M had a sidebar ad in the same issue, declaring Davis' version to be 'The Original! The Proven Monster!' and adding, as a threat addressed to Dolton, 'if they don't cool it, we'll cover "Shangri-La".' (See Figure 2.) 'Shangri-La' was another of Dolton's records which was climbing up the charts. Curiously, Dolton's version of 'Shangri-La' (recorded by Vic Dana) was itself a cover (of a version by Robert Maxwell). Given that the two ads appeared in the same issue, it is possible that the A&M/Dolton conflict was a bit of theater. Yet even as contrived drama it only makes sense with the presupposition that 'Fugitive' was a hit and that Dolton's 'Shangri-La' was the genuine article. The ads invite the reader to presuppose those things, against a background understanding that struggling to overtake a hit record with a cover is a losing proposition.

The shift away from jackal covers occurred somewhat later in foreign markets. Paolo Prato discusses songs from the 1960s that he knew growing

Figure 2: Two ads for 'Fugitive', from *Cashbox* 11 April 1964. Dolton's full-page ad (p. 39) and A&M's half-height sidebar (p. 36).

up in Italy. Although they were English or American pop hits, he thought of them as Italian songs because he only knew them from translated cover versions. Prato writes, 'Cover bands had an easy job in the 1960s, when many English and American records arrived in Italy: it was enough just to pick up a hit record and translate it to be successful' (2007: 458). He describes a drastic shift in the 1970s, though, both because more Italian musicians began to record original rock/pop songs and because Italian audiences began to expect English-language hits in their original versions. As a sign of similar shifts elsewhere: A Spanish producer in 1968 comments, 'Three years ago it was possible to get a Spanish group to cover a Beatles record and score a hit. But not any more. Spanish record buyers are demanding original versions and the language barrier has gone for good' (*Billboard* 1968).

Despite the trend away from it, however, jackal thinking did not end entirely. The music press notices periodically that there are covers that sell well. In 1965, Tom Noonan writes that 'cover disks are making it— that is, sharing the loot along with the big version.' Claiming that this is an exception, Noonan adds, 'In an earlier era, one version would generally step up and the others would drop out of the race' (1965: 1).

As albums became more widely available (rather than disks just being singles) new space for covers was created. When a song on an album was not released as a single, another artist might record a cover just to have it released as a single. Consider two examples. First, in 1961, 'His Latest Flame' was one of the tracks on an album by Del Shannon. Elvis Presley cut a version of the song just weeks later, and it was released as a single under the title '(Marie's the Name), His Latest Flame.' Elvis' single reached #4 on the *Billboard* Hot 100. Second, in 1968, 'Back in the U.S.S.R.' was one of the tracks on the Beatles' White Album. There were several cover versions, including a single by Chubby Checker just a few months after the album was released. Checker's single reached #82 on the Hot 100.

Even when many buyers started to demand original hits by original artists, some were still less discerning. Even though Woolworth's Embassy Records ended, low cost cover albums continued to be a thing. And the children's market allowed another niche for covers. A 1973 *Billboard* article comments, 'Unlike the pop field, where the buyer recognizes a cover immediately and, in fact, is usually looking for the original, the children's field is rife with cover records.' The article goes on to quote one record exec, who comments that cover records for children are 'good sellers' (1973a).

More recently, the shift to streaming music services has given a new place for jackal covers. Users search for a popular song but, depending on the exact search terms they use and where they click, may end up listening to a cover. Lizzie Plaugic writes, 'On platforms like Spotify, playing riffs on popular songs can lead to a far larger audience than recording original material — all you need is a song people are already searching for. … [W]ith a little creative track name optimization and a halfway decent recording, you could be looking at a potentially huge audience.' She adds a cautionary word which echoes the sentiments expressed in 1950s *Billboard* articles, but updated for the new technology: 'If streaming services are in fact the music-listening platforms of the future, expect a world with a few originals surrounded by dozens of copy and pastes' (2015).

To sum up: Despite changes in the music industry, there have continued to be covers in the sense that goes back to the 1950s. Changes in the market have discouraged covers in a certain respect but also created new opportunities for covering.

Before moving on, it is worth noting that charts and coverage in industry magazines were tools for commercial purposes and not a direct window into musical tastes. Rather than neutrally reporting trends, they could also reinforce and shape them. Elijah Wald notes that, by the late 1950s, there was a tremendous overlap between the *Billboard* pop and R&B top fifty. He suggests that this was due only in part to a convergence of musical style and that it also resulted from white teens starting to patronize the radio stations and record shops that served as the reporting basis of the R&B charts (2009: 180–181). There were subsequently changes in how the lists were constructed.

Even though there are limitations to what can be gleaned from industry magazines like *Billboard* and *Cashbox*, they nevertheless provide insight into the attitude toward covers at the time. What someone then actually wrote or how high a recording made it on the charts provides a check on present-day myths about what happened or biases about which records are significant. They are evidence, albeit imperfect, that my historical claims are not a just-so story that I contrived to underwrite my philosophical account.

Covers that hark back

Recall that Coyle claims covers, in a later sense of the word, are 'a way for performers to signify difference' and to 'project their identity.' He is right that this can happen, and his examples of early Elvis and the early Beatles are apt. Nevertheless, he is wrong that this necessarily involves the politics of race, and he is wrong that this is all there is to a 'cover record per se' (2002: 134).

First, regarding race: Consider two examples.

At the height of his fame in 1973, David Bowie released an album of covers. Instead of drawing songs from early American R&B, the *Pin Ups* album includes covers of songs which Bowie described as 'favorites from the 64–67 period of London' (Lenig 2010: 128). Stuart Lenig suggests that the album served 'as a crash course in British pop/mod culture of the last ten years' shaped to fit Bowie's vision (2010: 131). One might argue, because the rock bands of the 1960s that Bowie was covering were themselves influenced by black music and American R&B, that *Pin Ups* still figures in the complicated story of race and popular music in the 20th century. Those influences have nothing to do with *Pin Ups* as an album of covers, however. The identity which Bowie was forming and projecting was as a British rock star in relation to earlier British rock.

In the late 1980s, the punk band Social Distortion made a shift toward the subgenre of cowpunk— sort of the intersection of country music and punk. Their 1990 album included a cover of 'Ring of Fire.' It is typically understood as a cover of Johnny Cash, because Cash's version of the song is the classic despite not being the first published recording. Social Distortion's cover refigures it as a punk song, at once legitimizing the fusion of country and punk and projecting the group's identity as a cowpunk band. Lead singer Mike Ness recounts, 'This was during a period of time where there were a lot of those "what's punk versus what's not punk" discussions going on [but] I thought it was very punk rock to cover a Johnny Cash song' (Hodge 2017).

In both of these examples, an artist or band records a cover so as to establish their identity by their revision of and relation to an earlier version. Covers functioning in that way need not involve a dynamic of race, as it did when Elvis, the Beatles, or the Rolling Stones covered early R&B. Although race can arise as an issue in covers, that is because racial issues run deep in American culture and in American music. The phenomenon of the cover song is not essentially connected to it.

Second, regarding covers as the projection of identity: Although the last two examples were further illustrations of how musicians can use covers to establish their identity, there are many other reasons why musicians decide to cover.

There are periodic waves of nostalgia in popular culture. Lenig describes one such wave: 'Cover albums were rampant in the early seventies. During that time, Bryan Ferry, Bette Midler, Manhattan Transfer, the Band, Bob Dylan, Don McLean, were all engaged in cover projects. Television like *Happy Days*, plays like *Grease*, and films like *American Graffiti* celebrated a culture of past worship' (2010: 130–131). There was not a dominant new style, so 'cover albums were likely to at least rally some sales in a precarious and uncertain market' (2010: 131). In such a period, an artist might cover a track from the 1950s in order to craft their identity in relation to musicians of the earlier period. Yet they might do so simply in order to sell records. The same jackal thinking that justifies recording a cover of a current hit can, in the context of nostalgic market forces, justify recording a cover of a song from decades ago.

Yet, sometimes, there is no special reason for the timing of a cover. There just happens to be an earlier song which a new musician would like to play. Take 'Istanbul (Not Constantinople)', the example that started this chapter. It had the lyrical structure of a They Might Be Giants song already. Their songs have small stories and not-quite-serious drama, which they fold into what Jon Cummings describes as 'hyper-verbal alt-rock souffles' (Popdose 2011). So when they covered it, they were only projecting the identity that they had cultivated in all of their original songs. They were not trying to steal it, but neither were they using their relation to the original version to establish their bona fides. Nor was there any special reason why a song from 1953 should enjoy a resurgence in 1990.

Coyle is right that, as time went on, there were many covers of songs that had been hits years before (if they had been hits at all) rather than just covers of current hits. However, I think the explanation for this is rather more banal and superficial than he suggests.

In the 1950s, the history of recorded music was all relatively recent. In a series of *Billboard* articles in 1961, June Bundy writes about the *oldies* trend (1961a, b, c). There are two striking features of this trend.

First, although it included programming focussed on the big band era of the 30s and 40s, there was also a significant focus on late 50s rock and roll. Describing the success of one radio station, Bundy writes that 'the bulk of the programming is made up of hits from the '50s, [especially] r.&r. hits of

1955, 1956, 1957, and early 1958' (1961a: 1). From the standpoint of 1961, this reached back just a few years. As decades passed, that early rock and roll receded further. The oldies got older.

Second, as a result of the oldies trend some old songs had a resurgence of popularity. Bundy describes 'old hits' which had recently made it into the Billboard Hot 100. Old hits were competing with new productions; she explains, 'The new nostalgia trend isn't entirely to the liking of the recording industry, which is anxious to expose new wax product as well as old' (1961a: 47). In listing old songs making a resurgence, however, Bundy does not distinguish original recordings by the original artists from recent covers. Covers could be hits with songs that had been hit records before, and this means that it could make commercial sense to release covers of oldies. As the history of rock and roll grew from a few years to several decades, there were simply more earlier records to cover.

Because rock and roll began in the 1950s, it would have been impossible for a band in the 50s to cover a rock and roll song from an earlier decade. It was possible for bands in 1961 just because it was a new decade. Recorded music matured as a medium—it had been around for longer—so a musician who wanted to record an existing song was more likely to think of one that had already been recorded. Moreover, they were more likely to think of the original or classic recording of it as the source for the song. So it was more likely that their recording would end up being a cover.

This also explains why, over time, jackal covers make up a smaller proportion of all covers. Covering only threatens to hijack a hit when it is done just as the original is starting to climb the charts. A record is only a hit for a brief window of time, and a cover done later looks like returning to an old favorite rather than trying to steal sales from a hit. Dionne Warwick complains about other musicians recording covers that competed with her releases. Nevertheless, she admits, 'It's true that I've cut a lot of songs which other people did first but I always waited until the originals had had their chance. I had a hit with "I'll Never Fall In Love Again", but I had refused to record it until Ella Fitzgerald's original had died' (St. Pierre 1975).

The lessons of history

Where has this discussion of the history of the word 'cover' led us? The term first applied to new versions of songs which had already been recorded, where the original was already a hit or was expected to become a hit. It was extended to new versions of songs recorded longer ago. The motivation for publishing covers was often to capture sales which could otherwise have gone to the original. So when the original recording was by a black artist and the cover was by a white artist, covers could be a tool of exploitation and oppression. Yet covers were also sometimes recorded just because a current artist had fond regard for a classic recording.

This does help resolve some of the problems raised at the outset of the chapter. Let's take the problems of demo versions, songwriters, live versions, genre, and parodies in order.

1. Because cover songs began as a phenomenon of music publishing, for something to be a cover in the original sense there had to be an earlier published recording. Unpublished *demo versions* do not count. Robert Hazard's demo of 'Girls Just Want to Have Fun' is a peculiar case because it was published a few years later. Sonny West only recorded a demo of 'All My Love (Oh, Boy)', but his version of 'Rave On' was released as a single. Buddy Holly and the Crickets had hits with both, and they are usually mentioned together. If Hazard's demo had never been published and if it hadn't been for 'Rave On', then there might be no temptation to call Lauper's 'Girls Just Want to Have Fun' or the Crickets' 'Oh Boy' covers.

2. When a *songwriter* records their own song after someone else, it meets the simple definition of a cover. Yet they are not obviously driven by the jackal motive of stealing profits from the earlier recording, because the songwriter stands to gain royalties from the sale of either version. And they are not obviously establishing their identity by association with the song and the earlier recording, because as the songwriter they already were associated with it. As a result, the issue is vexed. It is tempting both to call it a cover and to not do so.

3. Although *live versions* can count as covers of earlier recordings, recordings cannot count as covers of earlier live versions. The former point is a natural extension of the concept. Cover bands play covers, even if they only ever perform live. The latter point, the requirement of an earlier recording, captures precisely the shift in the 1950s. The introduction of the term 'cover' coincided with a shift from live performances to recordings being the central musical commodity. Covers were a new thing, but playing songs that had been performed live by someone else was the usual condition of music since forever.

4.   Differences between *genres* are explained by the fact that the history of cover versions primarily centers on pop and rock music. Recording has played and continues to play a different role in jazz. Whereas the recorded track is the primary way to encounter a rock song, jazz recordings serve more as documentation of the original performance. When there are pop versions of jazz tunes or jazz versions of rock songs, it is unclear whether we should call them covers or not.

Genre difference also explains why recordings of classical music are not referred to as covers. Although aficionados might want recordings of specific performances, many buyers will take any recordings of a composition. Although I own a CD of Vivaldi's *The Four Seasons*, for example, I honestly have no idea which orchestra is recorded on it.

5.   In marketing terms, covers were distinguished from novelty records. This commercial difference may partly explain why *parodies* do not count as covers.

Admittedly, these considerations do not fully solve the problems. They neither provide a natural definition nor point to an essence. They do not reveal, out in the world, a kind to which the word 'cover' referred all along regardless of whether anyone could articulate its real definition. Both the word 'cover' and the practice of making covers serve a diverse range of purposes. Looking to history does not yield a forensic test which we can apply to determine whether, for example, Aretha Franklin's 'Let It Be' is a cover or not.

Covers and remakes

A number of scholars have used the broader notion of a *remake* to understand covers. In the next two sections, I consider two such approaches. The first aims to find a definition by way of an analogy with film remakes. The second distinguishes covers as a special kind of music remake, arguing for a narrower definition of 'cover' itself.

Covers and film remakes

Andrew Kania takes film remakes as an important clue to his understanding of covers (2006: 408–409, 2020: 237). Michael Rings takes this a step further, incorporating the analogy into a definition. He writes that a cover is 'a rock recording that captures a performance of a song that has already been recorded... by another artist and, as such, functions as a remake of the original recording, in a manner somewhat analogous to how remakes of films function in cinema' (2013: 56).

The greatest problem with such an approach is that it just replaces one puzzle with another. I do not know what to say about film remakes. It would take a great deal of time and thought for me to figure out what I ought to say, even tentatively. Moreover, this would be time poorly spent because film remakes are at best a loose analogy for cover songs.

Tony Kirschner points out several disanalogies. First, people encounter music in a greater variety of ways than they encounter films. He lists, 'live performance, recorded commodity, radio broadcast, music video, commercial jingle, movie and television soundtrack, and background music in public places' (1998: 249). Although a clip from a film can be used as part of a commercial or as a reaction to an internet post, music still shows up in a wider array of contexts.

Second, 'the lion's share of rock music production occurs at the amateur level' (1998: 249). Although computers and the internet have made video production easier, there is still a larger gap between what amateur video and what professional film look like than between how amateur rock and professional rock sound.

Third, music is more portable and travels more easily. Rock and pop music has influence around the world on local music scenes. Its influence is amplified by the fact that it shows up in more ways and impacts amateur practice.

I would add: Fourth, a song is simply much shorter than a movie. Listening to an original and a cover takes a few minutes, but watching an original film and a remake would take a whole evening. This difference changes how we can interact with them in fundamental ways, and is part of the reason that music appears in more contexts.

Fifth, movie remakes require large crews, from actors and directors down to gaffers and grips. Studio recording of a cover requires fewer people. Since a movie requires sound production and typically includes music, making it involves all the tasks involved in recording music *plus* all the visual tasks. In the limit, a cover requires just the musician. Although a film remake could in principle be made by just one person, that would be extraordinary. It is a common case for covers, though. It is not unusual for a cover played live or for an amateur YouTube video to not require anyone else. This explains why there is more amateur music production, and it greatly impacts the range of possibilities for cover versions.

Although Rings only relies on film remakes and covers being 'somewhat analogous', the connection is too weak to be helpful.

Covers and mere remakes

Theodore Gracyk argues that 'cover' has taken on a different meaning than it had in the early 1950s. Gracyk writes,

> Since the 1960s, the concept of the cover... normally refers to a communicative act of "covering." The cover record or performance is a version of an existing musical work. However, it is more than a version. It is a version that refers back to a particular performer's arrangement and interpretation of a particular song. (2012/3: 23–24)

He calls covers in the older sense *mere remakes*. He spells out the distinction in terms of whether the audience knows about— and is intended to consider— the earlier version. For a version to be a cover on Gracyk's account, 'A musician must intend to communicate with a particular audience — many of whom can be expected to recognize its status as a remake — and must intend to have the remake interpreted as referencing and replying to the earlier interpretation' (2012/3: 25). A version is a mere remake when the audience need not know about the earlier recording and is not intended to have it in mind. He writes, 'In contrast to a cover, a mere remake is a new recording of a song that is already known by means of one or more recordings, but where there is either no expectation of, or indifference about, the intended audience's knowledge of the earlier recording' (2012/3: 25).

His distinction is similar to Coyle's distinction between hijacking and recording a 'cover record per se' (2002: 134, discussed above). Gracyk, however, is not concerned with the projection of identity or entanglements with race. Instead, he thinks of a cover as involving reference to the earlier recording. The audience is invited to, even expected to, think of the new version in relation to the old one.

Many of the counterexamples to Coyle's thesis could be given again here as counterexamples to Gracyk's definition. He anticipates such a move, however. Precisely because the meaning of the term has shifted, the past will be rife with apparent counterexamples to his claim about the current meaning. Gracyk writes, 'The analysis offered here does not pretend to capture all uses of "cover" in recent popular music. Concepts evolve, and therefore the early uses of a term are not an infallible guide to its present meaning' (2012/3: 25).

There are limits to this maneuver, however. On Gracyk's account, the 1987 version of 'I Think We're Alone Now' by Tiffany is merely a remake of Tommy James and the Shondells' 1967 hit. It is not a cover, he says, because there was no expectation that Tiffany's adolescent fans would recognize it as a remake (2012/3: 25). Nevertheless, a news report in 2019 says, 'Tiffany Darwish, known as Tiffany, is today possibly best known for her Billboard No. 1 cover version of the song "I Think We're Alone Now"' (DiGangi 2019). And Tommy James himself sees it as a cover, recounting in an interview that 'Tiffany came up to me at a convention to apologise for covering us. I said: "Are you nuts? I should be thanking you." She did a great job…' (Simpson 2019). These are recent choices of words, and cannot be dismissed as a relic of mid-20th-century usage. One could argue that James is wrong to describe Tiffany's version as a cover, but saying that the usage is mistaken is different than saying it is only evidence of earlier meaning.

Consider also that Tiffany released a new version of the song in 2019. It would be awkward to say that it is a cover of her 1987 version, because Tiffany is the artist who recorded both. It might be another cover of the 1967 original. Some news coverage avoids calling it a cover, calling it a remake instead, although one report does describe it as 'a more crunchy and rock-inspired cover' (DiGangi 2019). Regardless, her new interpretation of the song is explicitly in relation to her earlier version. Many listeners will remember the song from the 1980s, she knows this, and she intends for them to think of it. Many of her old fans will both remember the earlier version and be able to share the new version with their teenage children. So Tiffany's 2019 version counts as a cover of her 1987 version under Gracyk's definition.

Gracyk's definition gets it wrong in both cases. The 1987 version of the song counts as a cover by contemporary common usage, but it does not meet Gracyk's definition. The 2019 version is not obviously a cover— at least not of the 1987 version— but it meets Gracyk's definition.

What are we doing here?

Part of the difficulty in defining 'cover' is that it is unclear what kind of answer we want when we ask what a cover is.

First, we might be trying to give an *analysis*. In providing a conceptual analysis, a philosopher takes the concept to be pretty well settled. The philosopher's task is just to make explicit what we already understand. Traditionally, this has meant to provide a precise definition— necessary and sufficient conditions. An alternative is to analyze the concept as a property cluster. Regardless, analysis starts from common usage and is taken to be successful if it accords as much as possible with common usage. Because of that, it becomes a relentless cycle of proposal and counterexample. For every proposed definition, the philosopher stumbles on something that counts as a cover that is left out of the definition or something that should not count but is left in. A revised definition which handles those cases is devised, but counterexamples to the new definition arise. The process repeats.

Second, we might be trying to provide an *explication*. Instead of trying to say what we already mean by a term, explication is aimed at figuring out what we *ought* to mean. The term 'explication' comes from Rudolf Carnap, but the approach has more recently been called *conceptual engineering* (Cappelen 2018). Instead of abiding with the messy, imprecise concept we already have, the conceptual engineer tinkers with it so as to produce one that is more fit to use. Explication is not vulnerable to counterexamples in the way that analysis is. What matters is not whether the engineered concept fits with common usage but whether it can fulfill the function of the original term. It is often the case that the more useful concept will require reclassifying some specific cases.

Explication is often a good strategy in scientific work. Carnap gives the example of *fish* (1962: 5–6). Before the 18th century, it was standard to count whales as fish. Carl Linnaeus' taxonomic system counted whales not as fish but as mammals. Although this required revising common usage, it better tracked distinctions that biologists were concerned with making. Now modern evolutionary systematics has put pressure on the Linnaean concept of fish. (Another example, which Carnap develops in his own work, is *probability*.)

Explication need not always be scientific. However, it requires that the context determine what function the concept is supposed to fulfill (Nado 2021). The problem with the *cover* concept is that it has no clear function. It is possible to propose definitions which would be more precise and give definite answers, to say about any recording whether or not it is a cover, but our interests and practices provide too little constraint to make any such explication the right one.

An artificially precise definition risks stipulating answers to substantive questions. For example, we might start out with the question of whether covering is an essentially racist practice. If we accepted Don McLean's definition of a cover as a racist tool, then we would say that it is racist— but this result would follow just from the definition itself.

One might argue that the function of the *cover* concept is to help understand musical practice, and surely that is right. However, which musical practice should it help us to understand? The answer cannot be *covering*, because the function is meant to help us say what counts as covering. If we were concerned primarily with how musicians can use old songs to convey messages to new audiences, then something like Coyle's or Gracyk's definition might work as an explication. Yet this would leave out other important functions of 'cover' talk. Recall the fans from the outset of the chapter surprised to discover that TMBG did not write 'Istanbul (Not Constantinople).' Recall Sandy Brown lamenting the 'jackal thinking behind cover versions.' I would like to offer an account that addresses their concerns, too.

Kurt Mosser argues that '"cover" is a systematically ambiguous term' corresponding to a cluster concept (2008). I would push the matter even further. There is not a clear enough cover concept to make analysis worthwhile, and our talk of covers does not have a unified enough function to make explication worthwhile.

Of course, there are things we can say in general about covers. Covers, in the sense that I am concerned with, are musical performances or recordings that stand in relation to an earlier recording. This does impose some restrictions. Where there is no earlier recording, I will not count a version as a cover.

Moreover, the word 'cover' has extended metaphorical uses which I do not have in mind. Take two examples: First, Greg Metcalf discusses covers not just of songs but of artists, as when a performer adopts something like the persona of Bob Dylan without actually playing any of Dylan's songs

(2010: 183). Second, a character in the television show *Lucifer* offers praise for a copycat using the same M.O. as an earlier serial killer by saying, 'This new guy is not the original, okay? But he is a damn sweet cover' (2018). Without a definition of 'cover', I cannot say that these uses are somehow illegitimate. However, I can say that they do not describe the kind of covers that this book is about.

This yields a necessary condition— that to be a *cover* something needs to be a musical version which covers an earlier recording. One might think this puts us right back at the definition proffered at the beginning of the chapter, namely of a cover as a 'recording of a song that has been recorded previously by another artist' (Zak 2001: 222). I do not think so.

First, a necessary condition is not on its own a definition. There are, as we have seen, several ways in which a version can meet the condition but still not be a cover.

Second, and more contentiously, I am not sure that the cover and the earlier recording will always be the *same* song. Covers can be different than the original, and maybe they can be so different as to be a different song. The necessary condition allows for this possibility, whereas the usual definition does not. (This point is addressed further in Chapter 5.)

Regardless, the usual definition allows us to identify a typical case of a cover. There are lots of extraordinary cases of versions that meet the definition as written but are not covers and maybe some that are covers even though they fail to meet the definition. We could try different formulations in hopes of hitting on a definition of 'cover', either to capture the existing meaning (analysis) or to craft the term for some specific function (explication). My suggestion instead is that we should forge ahead. There are interesting things to be said about covers, there are important distinctions to be made, and we can do that without a precise definition.

Conclusion

Carnap reflects on his attempts to give an explication of *philosophy*: 'Yet actually none of my explications seemed fully satisfactory to me even when I proposed them… Finally, I gave up the search.' He advises that 'it is unwise to attempt such an explication' because it is better to work on philosophical problems than to refine a sense of 'philosophy' that precisely fits them (1963: 862).

I suggest that what holds for 'philosophy' also holds for 'cover.' Common usage identifies some recordings as *covers*, and I take that as given. A cover is pretty much whatever audiences and critics count as a cover, but that alone is of little consequence. Considering those so-called covers, there are philosophical questions that arise and useful distinctions to be made. I start that work in the next chapter.

2. Kinds, Covers, and Kinds of Covers

At the end of the last chapter, I refused to give a definition of 'cover' but promised that there were other important distinctions to make. In this chapter, I distinguish songs, performances, and tracks— arguing that all three are objects worthy of appreciative attention. I then make a distinction between two kinds of covers: mimic covers and rendition covers.*

Musical works

We can easily distinguish between a song itself and performances of that song. For example, the song 'Happy Birthday' can be sung by different people on the occasion of different birthdays. When it is sung on a particular occasion, there is one performance of it. There may be many different performances simultaneously, in different places around the world, to celebrate different people's birthdays. All these different performances are performances of one and the same song.

Some recordings are just intended to provide access, as much as possible, to an original performance. For jazz records, the entire ensemble typically plays together. The recording is taken from just one performance rather than multiple performances mixed together. Often, the best take is chosen for release. In some cases, multiple takes are released— as on the Miles Davis album *Bag's Groove*, which begins with two takes of the title tune. Jazz listeners want to hear the performance, and the recording is expected to transmit that as much as possible.

In contrast, recordings in rock and pop music typically do not correspond to any single performance. Different musicians record their parts sep-

---

*Parts of this chapter are adapted from work I coauthored with Cristyn Magnus and Christy Mag Uidhir (Magnus et al. 2013).

 https://doi.org/10.11647/OBP.0293.02

arately, rather than together. For each part, the final track includes material from multiple separate performances. Sometimes these are performances that could not have occurred together, as when Stevie Wonder recorded four different instrumental parts and vocals for 'Superstition.' Not only are the best bits selected and assembled, they may also be transformed by various effects. An egregious example of this is the heavy use of Auto-Tune, software originally introduced in the late 1990s, which makes sounds match a prescribed pitch but can also make them sound artificial and robotic. Furthermore, samples and sounds can be introduced in production that do not correspond to sounds which were ever made by performers.

Theodore Gracyk notes that a track which is the result of editing and post-production in this way 'is accessible only by playing, on machines, end-products derived from a montage of partial performances' (1996: 34). For rock music, the recorded track is the way that most people encounter the song. It is what they hear on the radio and (depending on the time period) on records, tapes, CDs, or streaming internet audio. As Gracyk puts it, tracks 'are the standard end-products and signifiers in rock music' (1996: 36).

What he has in mind is the master track or copies of the master. Talking about recording and production, the word 'track' has another use. Different parts or sets of sounds are recorded to different tracks. When recording is done to tape, the tracks are literally paths on the tape itself. With computers, they are separate sound files. These are mixed together for the final version. When the contrast is not at issue, the final version is simply called a track—as when discussing the tracks on a published album. I will not have much cause to talk about separate recorded parts, so I will use 'track' to mean the finished product.

This makes for three different kinds of things:

1. Lyrics and musical structure provide the identity conditions for the *song*. This allows for different interpretations and arrangements, so that performances and recordings of the same song can sound rather different.

2. A *performance* is an event which happens only once. Although we can listen to a recording of it later, the same performance cannot happen again. A repeat performance, even if it sounds the same, is a separate instance of the same song.

3. A *track* is a specification of sounds. It can exist as multiple copies in different media, and it can be played back repeatedly.

Note that these are explications in the sense discussed in the previous chapter. Making the distinctions in this way allows me to precisely discuss particular issues, but it does not perfectly match ordinary usage. It is common to refer to a cover as a 'cover song' even though a cover is a performance or recording rather than a song in the sense distinguished above.

Contrast the current Apple Music Style Guide, which defines 'Song' as 'An audio recording' and 'Track' as 'A song or music video' (2021). The Spotify Metadata Style Guide is similar. These definitions do not fit common usage, either. Instead, they are specifications for the purpose of content metadata. Like mine, these definitions are explications. It is just that their purposes are different than mine. Apple and Spotify want to maintain enormous digital storehouses of media files, whereas I want to understand covers.

Even experts will often not mark the distinction between what I am calling songs and tracks. Producer and educator Rick Beato comments in a YouTube video discussing a recent hit, 'That's a very well produced song. It's got a great melody. Interesting twist in the chord change. Whoever mixed it, the low end is massive' (2021). The melody and chord changes are features of the songwriting, but the mix and the low end are features of the recording. Similarly, a list of '500 Greatest Songs of All Time' in *Rolling Stone* slips from high-level considerations of lyrics and melody to low-level considerations of timbre and production (2021). The point is not that Beato or the writers at *Rolling Stone* would not recognize the distinction between songs and tracks, but rather that they do not always mark the distinction in reacting to music.

Admittedly, none of these things would show up on a list of what fundamentally exists. Neither songs, performances, nor tracks are mentioned in physicists' Grand Unified Theory— nor do they appear on philosophers' lists of fundamental entities or categories. Nevertheless, songs, performances, and tracks all exist in an ordinary sense of *exist*.

The boundaries of each category will not be perfectly precise, however, and the process of writing a song, performing it, and constructing a track can merge together. When a musician noodles around on an instrument, toying with a riff, there is no sharp divide between the playing and the songwriting. If they record themselves doing that, then there may be no sharp line when songwriting ends and constructing the track begins.

Albin Zak discusses the example of the Beatles' 'Strawberry Fields Forever.' It was originally introduced to the band with only guitar and vocals. A version was then arranged and recorded with the whole band. Setting that version aside, producer George Martin wrote scores for cello and trumpet parts, which were recorded along with additional percussion parts. The tape of the percussion instruments was reversed, making a kind of sucking sound. The resulting version is 'a layering of nonsynchronous performances through overdubbing, and, in the backwards percussion, timbral characters produced by the recording medium itself.' Ultimately, the decision was made to use both versions. By slowing down one and speeding up the other, the pitch and tempos were shifted 'to make a credible, if not altogether unobtrusive, edit' (2001: 36). The resulting track is the band version for about a minute and then shifts to the more complicated second version. In cases like this the process of composition is interwoven with the process of recording.

Tracks

Even recognizing that all three things exist, it is natural to wonder if they are equally important. Some philosophers have argued that tracks are what *really* matter in rock music.

Gracyk claims that 'rock music is not essentially a performing art, however much time rock musicians spend practicing on their instruments or playing live. And while I do not say that it is essentially a recording art, I do contend that recording is the most characteristic medium of rock' (1996: 75). Gracyk points out that many rock musicians have been especially interested in creating complex sounds in the studio— that is, in constructing tracks. For example, he quotes John Mellencamp who likes the creativity of crafting songs in the studio but sees stage performance as repetitive (1996: 81). Gracyk notes that this interest in studio craft has led to some rock musicians giving up performance entirely. He writes, 'Some rock musicians, like the Beatles and Steely Dan, eventually abandon the stage without harming their careers' (1996: 81). Most musicians do not have the luxury that the Beatles and Steely Dan had. If Gracyk is right, though, then touring and performance are just economic rather than artistic necessities.

Andrew Kania writes similarly that 'rock musicians primarily construct *tracks*' and thus that tracks 'are at the center of rock as an art form' (2006:

404). Yet Kania presses the point further. Because a track is not a performance, a rock song is not a work for performance the way that (for example) a classical symphony is. Rather, the rock song is just what is manifest in the track. The track itself is the work. Thus, Kania argues, 'rock songs are not works, nor are they *for* anything in particular' (2006: 404). Although a rock song can be performed live, it is just a loose specification of a sound that is only fully realized on the track. Kania writes, 'when listening to a rock track, one does not focus on the thin song manifested in it, nor wonder what another rock band would have done with it; rather, one listens to the track as an entity complete in itself' (2006: 409). This would give us a different reading of the quotation from Rick Beato discussed above— instead of mixing song and track evaluation, Beato is taking all the features to just be features of the track. To put Kania's claim in Gracyk's terms, he insists that rock is essentially a recording art. The record is the work.

The claim that tracks are *the* important artistic product in rock music minimizes the status of both performances and songs. Michael Rings, who endorses the view, makes that consequence clear when he writes that 'the central work in rock music is the sound structure captured by the recording… as opposed to a song, a score, or a performance (all of which may function as works in other musical traditions)' (2013: 56).

Let's take a closer look at performances and songs.

Performances

Even though many rock and pop musicians focus primarily or even exclusively on making tracks, that does not mean performances are never events worthy of aesthetic and artistic appreciation. Most musicians start out performing before they start recording tracks. The fact that tracks can be constructed on a laptop computer in a musician's bedroom means that some musicians may now jump straight to making tracks, but that is only a recent development. Treating performance as not a work of rock would mean that bands who had not recorded yet had never made rock music— that they had at most done some kind of warm up activity for the real thing.

Moreover, there are subgenres in which performance is the primary focus and where recordings are more documentary than they are constructed tracks.

First, David Goldblatt gives the example of urban vocal groups of the 1950s and 60s. In doo-wop music, he argues, 'the creation of singers, as well as songs, their rehearsing and performing, was a significant part of the historical process of rock without which many singers and songs never would have made it to a point where they could be packaged and distributed'—that is, singing on street corners necessarily preceded recording tracks (2013: 102). Although a great many black vocal groups started recording in the 1950s— Goldblatt echoes Bill Millar's estimate of 15,000 such groups— most were ensembles who came to the studio already having performed together (2013: 101).

Second, Christopher Bartel (2017) gives the example of hardcore punk in the early 1980s. Hardcore bands focussed primarily on live performance, and records were more documentary than an occasion to craft a distinct, original sound. Mike Watt of the band the Minutemen says, 'It was the exact opposite of the big leagues — we didn't tour to promote records, we put out records to promote tours. Records were like flyers...' (Blush 2010).

Third, there are jam bands. Bartel writes, 'Like jazz musicians, jam bands place greater value on improvisation rather than album fidelity while yet remaining broadly within the tradition of rock' (2017: 152). The Grateful Dead both performed extensively and made no effort to make their live sound echo their studio recordings. Fans of the band show great interest in unofficial live recordings, and the band supported this interest. As of this writing, the Internet Archive hosts 15,986 recordings of Grateful Dead performances. To take one other example, the Dave Matthews band produces studio tracks and their performances on television basically follow the recorded versions. In concert, however, they perform as a jam band—improvising and performing songs differently at each concert. This practice shows that they and their audiences care about performance.

Songs

The genre of doo-wop also provides an argument for the importance of songs. More than just a matter of performance, doo-wop exhibits a distinctive kind of songwriting. In the Five Satins' 1956 hit 'In the Still of the Night', Goldblatt points out that the 'shoo doot 'n shoo be doo' is part of band member Fred Parris' composition. The nonsense syllables serve as 'an instant phonetic souvenir easily sung by most anyone.' Unlike scat singing in jazz, which involves improvising nonsense sounds, in doo-wop the sounds are 'built into the fabric of the songs' (2013: 103). His discussion of the use of nonsense syllables and repetition in doo-wop requires considering the songs themselves.

Songs are also important objects of appreciation in rock and pop music more broadly. As Franklin Bruno notes, 'a number of rock musicians are frequently credited with excellence or merit as songwriters by listeners and practitioners' (2013: 67). Although Bob Dylan is a singer/songwriter who figures as a frequent example in Gracyk's discussion of tracks, Dylan's songs are worthy of attention in their own right. When Dylan was awarded a 2008 Pulitzer Prize, it was for his 'lyrical compositions of extraordinary poetic power'— similarly for his 2016 Nobel Prize in Literature.

Gracyk, who focusses on musical features rather than on lyrics, would take the invocation of Dylan's prizes to be unconvincing. He writes, 'To be blunt, in rock music lyrics don't matter very much. Or, to be more precise, they are of limited interest on the printed page, divorced from music' (1996: 65). Lyrics, he notes, are often written after the instrumental parts are mostly complete. For myself, I care more about lyrics than Gracyk does. In general, different listeners may respond to different things— and the same listener may respond to different things in different cases. When listening to Dylan's 1964 'My Back Pages', I find the song to be more rewarding than the full sound of the track. For Dylan's 1965 'Like A Rolling Stone', the opposite. (I won't stake my life on these assessments, and there is more about Dylan in the next chapter.)

Of course, we usually consider rock songs by comparing renditions rather than by looking at sheet music or written lyrics. Nevertheless, we can see the various versions as expressing the possibilities of the song. As Bruno notes, 'songs can be judged excellent by listeners who do not find merit or take pleasure in their best-known renditions' (2013: 67). Someone may enjoy a cover despite disliking the original, on account of how the same song is handled in the two tracks. Conversely, covers which fall short of the original can reveal the complexity of a song— where it seemed simple in the original, that was only because of the great skill involved in its recording and production.

Part of the fascination of covers is that they serve as a laboratory to reveal the limits and potential of a song in this way. Even for a great song which is originally recorded in a great track, the cover can provide a different perspective. Paul Dempsey of the band Something for Kate comments that their goal with a cover is to let the audience 'hear something differently and see how great it is when a really well-written song can be interpreted in a different way and still stand up' (JJJ 2021b).

The slipperiness of 'work'

Where does this leave the question of which of these things is the musical *work*? The everyday concept and phrase, 'the work', is woefully slippery. Discussions of popular music often do not use the word (Horn 2000). Let's consider three things it might mean.

First, looking at what it means to be a *musical work* in classical music, we can take a work to be the full specification of what a musical event should sound like. The work, a composition as written in a score, specifies exactly what notes the performer is supposed to play. In rock and pop music, the full specification of the sound is given by the track. This yields the conclusion that the track is the work.

As Lydia Goehr argues, understanding works and performances in this sense just reifies 'ideals that exist within classical music practice' (2007: 99). Since about 1800, classical music has been guided by an ideal of 'perfect compliance'— that the performer should play all and only the notes in the score. Goehr points out that this norm is a distinctive development. She writes, 'it is significantly this ideal that serves to distinguish the practice of producing performances of classical music works from the performance practices associated with other kinds of music' (2007: 99). Although it is typical to hear rock and pop songs as tracks, the discussion of performances and songs above shows that tracks do not (or at least do not always) play the same regulative role that compositions do in classical practice. So this sense of 'work' will not underwrite an argument about rock music in general.

Second, we can take the work to be the durable object. This marks the distinction between works (as repeatable) and performances (as fleeting). Without recording, each performance is a one-time event while the work could be performed many times. The score for the work is a tangible thing. It is a product or commodity which can be stocked on a store shelf, held as inventory, and sold. With recording, the record is the tangible thing. Even though digital files have replaced physical media, the files too can be stored and traded.

This is a difference in the metaphysics, what kind of thing each of them is. Yet it has consequences for our appreciative practices. The fact that an un-recorded performance is a singular event whereas a track may be listened to many times makes a difference in listening (Davies 2001). However, a recording of a performance might give a listener access to all the appreciatively relevant features and can also be listened to repeatedly (Mag Uidhir 2007, Magnus 2008). So this metaphysical distinction does not establish the exclusive primacy of tracks.

Third, we can take the work to be the object of artistic appreciation—as Charles Nussbaum puts it, 'the object of proper musical regard' (2021: 329). Gracyk points to 'the *works* that the artist *sanctions* as items for appreciation and critical evaluation', thus distinguishing works from sketches and preparatory drafts (1996: 35). Because (many) musicians intend for audiences to primarily appreciate their published tracks, this yields the conclusion that the track is the primary work in rock music.

As examples of doo-wop and punk show, however, not all rock musicians take tracks to be their primary focus. Even if they did, it is not clear that such an intention would proscribe appreciating a song or performance.

Imagine a luthier who is very proud of the guitars they make. They record themselves playing one of their guitars. Because it is intended to help them sell guitars, the recording is meant to be appreciated and evaluated as documentation of how their guitars sound. Even so, I do not see anything untoward about a listener appreciating the performance itself. You cannot assume that the guitar will sound that good if you buy one and play it yourself, but that just underscores that appreciating the performance— although permissible— is different than the intended appreciation of the instrument.

In much the same way, people do appreciate songs and performances even from artists who see tracks as their primary product. And there is nothing wrong with that.

Ultimately, the question of which of the three is *the* work is misconceived. They are all important parts of musical practice, both for artists and audiences. As Bartel suggests, 'we should see rock as a tradition that has three activities at its core: songwriting, live performance, and track construction' (2017: 153). He calls this view 'rock as a three-value tradition', and Dan Burkett calls it 'a pluralist ontology of rock' (2015).

One might object to this pluralism on grounds of parsimony or simplicity. The idea is that Ockham's razor should stop us from positing many works rather than one. Ockham's razor is the principle that entities should not be posited any more than necessary, but the objection would misuse the principle in at least two respects. First, recognizing all three categories does not require recognizing any extra entities. Songs, performances, and tracks all *exist*, regardless of which are works. Second, the fact that important parts of musical and critical practice look to songs and performances means that recognizing them does not multiply distinctions beyond necessity. Rather, it is exactly what is necessary to make sense of these practices.

Terminology

Gracyk and others are right to insist that tracks are importantly different works than songs. Especially before Gracyk, authors often failed to distinguish them. So pluralism acknowledges Gracyk's greatest contribution.

As I noted earlier, distinguishing these three kinds of things as *songs*, *performances*, and *tracks* is an explication. Since they mark useful distinctions, I will follow that usage as much as possible in this book. I will use the word 'song' to indicate the repeatable thing which loosely specifies what the lyrics and tune should be, the word 'performance' to indicate either the live event of playing a song or a recording that documents such an event, and the word 'track' to mean a recording constructed in the studio.

A few other terms will also be useful. When the distinction between a recorded performance and a track is not relevant, I will use the word 'recording.' When the distinction between performances and tracks is not especially relevant, I will use the word 'version' to mean a performance or recording of a song.

What is it that a cover is a cover of?

The explications above make more precise the condition (from the end of the last chapter) that a cover must be a musical *version* which covers an earlier *recording*. Covering is a historical relation between an earlier recording and a later version of a song. The earlier version is taken as the canonical one, and the cover is based on it. It is the target of the cover, the source for the new version. Kurt Mosser calls it the 'base' version, the one 'that, due to its status, popularity, or possibly other reasons, is taken to be paradigmatic' (2008).

The target of the cover is often the original or first-published recording of the song, but not always. For example, the Beatles' version of 'Twist and Shout' was chosen by *Rolling Stone* readers as one of the greatest cover songs of all time in a 2011 poll. In that article and in many other places, it is described as a cover of the Isley Brothers' 1962 track even though the first release of the song was by the Top Notes in 1961.

Kania argues that covers are not indexed to earlier recordings in this way. He argues instead that 'covers can be grouped together as tracks intended to manifest the same song' (2006: 410). For example, we can organize the Beatles' track, the Isley Brothers', the Top Notes', and dozens of others besides as versions of 'Twist and Shout.' Although it is true that we can group them together that way, it leaves out important information about inspiration and influence. Calling the Beatles' version a cover of the Isley Brothers suggests that they had it in mind. The sound of the Beatles' track reflects their choices and abilities in relation to the version that inspired them. We can credit the cover for improving on the original or fault it for falling short. Contrawise, it would make no sense to think of the original as influenced by the cover. Just considering all the versions of 'Twist and Shout' together would overlook this asymmetry.

Still, we should consider the two reasons Kania offers to support his claim that a cover is indexed to a song rather than to an earlier version.

First, Kania maintains that appreciating a cover does not require considering the earlier track of which it is a cover. He writes, 'although a band may take just one version of a song as their target, knowledge of this does not seem relevant to critical assessment of their track' (2006: 411). He argues for this by example, providing an extended discussion of the Pet Shop Boys' 1987 hit cover of 'Always on My Mind'— a song made famous by Elvis Presley in 1972. Kania writes, 'Willie Nelson covered "Always on My Mind" in 1982, between the Elvis and Pet Shop Boys versions. Both of the later versions are covers of the same song. It would not make any difference to this situation if the Pet Shop Boys had never heard Elvis's track and only intended to cover Nelson's' (2005: 411). Discussing the Pet Shop Boys' cover, Kania notes that the chord progressions and arrangement are more complex than in Elvis' version. By their own declaration, 'the Pet Shop Boys wanted to construct a track that sounded as different from Elvis's as possible' (2006: 410). So it is unsurprising that critics and audiences evaluate the cover in relation to Elvis' track, rather than (as Kania would suggest) just evaluating it as a version of the song. For example, *The Telegraph* eulogizes that the cover 'elevated Elvis's tender elegy… into a monumental explosion of high pop camp' (2004).

Also consider Bob Dylan's 1967 'All Along the Watchtower.' It was covered the following year by Jimi Hendrix, and the musical changes introduced in Hendrix's version became iconic. As Janet Gezari and Charles Hartman write, Dylan 'adopted Hendrix's stylistic take on his song, as revealed in many live recordings in the seventies and after. In effect, he covered a cover of his own song...' (2010: 165). To understand what Gezari and Hartman as saying, one needs to think of Dylan's later versions not merely as instances of the song (which he wrote) but as being in relation to a version which they are covering (that is, Hendrix's recording).

In both of these cases, critical assessment of the cover turns on considering it in relation to an earlier track taken as a canonical original. However, in neither case is the canonical original the first recording of the song— Gwen McRae and Brenda Lee both released recordings of 'Always on My Mind' before Elvis, and Dylan himself had first released 'All Along the Watchtower.'

Second, Kania writes that 'rock audiences seem to group covers with respect to the song they are intended to manifest, rather than simply by the track(s) taken as the immediate object of the covering intention. Covers do not only occur paired one-to-one with originals' (2006: 411). He is right that the covering relation— although typically a one-to-one relation between a cover and an original— is not always so simple.

The case of 'Respect' illustrates how the relation can be history-relative and audience-relative. Aretha Franklin's 1967 hit version was a cover of an Otis Redding hit from two years prior. Franklin changes around the lyrics and narrative of Redding's song, and some of the most memorable lyrics are Franklin's addition. She adds spelling out the word 'R-E-S-P-E-C-T' and saying 'Find out what it means to me.' Although Franklin's version was not the first, it can be taken as canonical and as the target of a cover. Gracyk writes, 'It's easy to forget that Aretha Franklin's titanic, defiant version of "Respect" (1967) is a cover version of one of Otis Redding's 1965 hits, and her action of appropriating "his" song was originally an important element of its affective power' (2001: 209). Kelly Clarkson performed 'Respect' in the first season of American Idol in 2002, and Clarkson's version is labelled as an Aretha Franklin cover. We can only understand this shift by thinking of a cover as *being of* a track which is taken as canonical, rather than as being merely another instance of the same song. Clarkson takes Franklin's version as canonical both by having it in mind and by singing Franklin's additional lyrics. We would evaluate Clarkson's version differently if— impossibly—

she had never heard Franklin's version but had just made up for herself the lyrics which differed from Redding's.

Moreover, the relation can be one-to-many instead of one-to-one. Consider college a cappella choir versions of Dr. Dre's 'Bitches Ain't Shit'; for example, by DeCadence at UC Berkeley in 2010 and Continuum at Eastern Michigan University in 2011. The song originally appeared on Dre's 1992 album *The Chronic*, but Ben Folds had a hit cover of it in 2005. Folds' cover is a piano and voice arrangement that transforms Dre's rap into a serenade, and he has sometimes performed it a cappella. The various vocal groups who have performed it follow Folds' version, adapting it for choir. Audiences were likely to be familiar with both the Dre and Folds versions, so we might sensibly say that the choir performances are covers of *both* hit tracks.

So the relation can be more complicated than just pointing to one earlier track which counts as the canonical original for all time. Nevertheless, there is always a musical-historical relation that connects a cover to one or more earlier recordings. Absent that relation, it is not a cover at all but just a version of the song in question.

Mimics and renditions

Of course, there are different possible relations that a cover can have to its target original.

Some covers aim to echo the interpretation and arrangement of the canonical recording precisely. Call these *mimic* covers. A perfect mimic cover would be indistinguishable from the original.

This means that a recorded mimic cover is, in a sense, redundant. One could always just listen to the original instead. Even though one can easily go listen to the original track, it is not so easy to attend a performance by the original artist. A cover band might be the only option. The result is that most mimic covers are performed live. Mosser presses this further, writing that 'these kinds of covers are invariably live performances' (2008: §II.a).

However, there are recordings of mimic covers. Consider two kinds of examples.

First, mimic covers can be technical exercises. Zak describes a mimic cover recording by Karl Wallinger: 'Much as a composer of scores might copy out a favorite piece by another composer, Wallinger remade "Penny Lane" note for note and sound for sound, including reverb.' The point is not that anyone else would want to listen to Wallinger's mimic track. Rather, Zak explains, 'What he learns from such an undertaking he uses in making his own records, which in turn bear the sonic influence of, among other things, the later Beatles albums he is partial to' (2001: 36). This technical exercise is like a jazz student transcribing and performing a mimic of a famous jazz solo. The point of such an exercise is not to entertain an audience, but instead to develop skills and a vocabulary of sounds that can be used later in original undertakings.

Second, mimic covers can be jackal tracks meant to cash in on the success of the original. The Typhoons were the house band at Embassy Records that recorded the label's mimic covers of Beatles' tracks. They covered dozens of Beatles songs from 1962 to 1965, but did nothing outside the studio. Somewhat later, from 1968 to 1985, there was a record series called *Top of the Pops* produced by Hallmark. Each of the albums in the series was a collection of recent songs recorded by in-house artists and sold at a low price. In the mid-1970s, one of these even topped the best-selling album charts. People did not buy it for those versions especially, but because it was cheap and readily available. (The rules for what was counted in the charts were changed so that this did not happen again.) Later still, there are artists who post mimic covers to electronic music services in hopes of profiting from accidental clicks. All these mimic covers are meant for an audience, although an audience that only listens to them for reasons of economy or misdirection.

As Gaynor Jones and Jay Rahn suggest, the creation of mimic covers is 'closer to classical performance practices than to those which we usually associate with popular music' (1977: 85). A pianist playing a Beethoven piano concerto typically tries to play what is written in the score, note for note. In a similar way, someone performing a mimic cover takes the original track as specifying what a version of the song ought to sound like. They attempt to follow the original track, sound for sound. In a certain respect, this is even more restrictive than the practice in classical music. Practices in classical music allow 'narrow decisions of dynamics, phrasing and articulation' (Hamm 1994: 149). As Zak notes, there has been a trend of narrowing these decisions even further; he writes that 'classical compositions of the nineteenth and twentieth centuries have generally tended toward increasingly detailed written specifications, which attempt to define a work's parameters with ever greater precision' (2001: 42). This is what Goehr (2007) calls

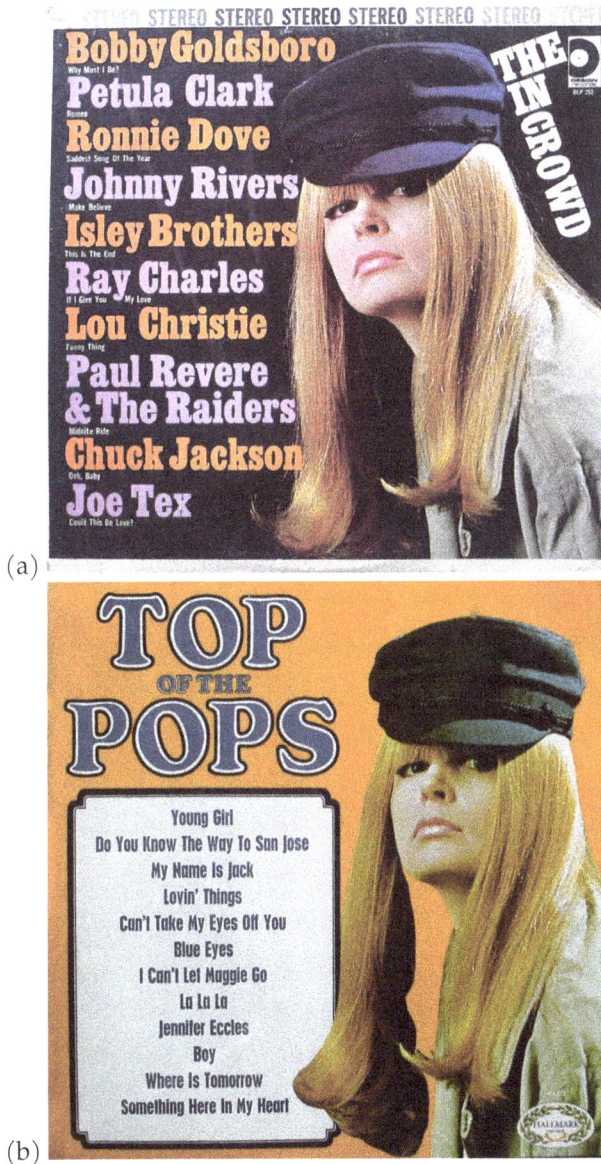

(a)

(b)

Figure 3: The *Top of the Pops* record series comprised albums of mimic covers put together on the cheap. Even the album art was recycled. (a) A 1966 publisher sampler album from the U.S. (b) The first *Top of the Pops* album, released in 1968, reuses the same photograph. It was used again as part of the album cover for another *Top of the Pops* release the next year.
Images used with permission of the Pickwick Group.

the ideal of 'perfect compliance.' Tracks, 'like scores, are detailed and fixed representations of musical thought' (Zak 2001: 42).

Other covers do not treat the original in this way and do not aim simply to echo features of the original. Call these *rendition* covers. Most recorded covers are renditions, where the recording artist or band is not attempting to fully emulate or impersonate the original. Some renditions are fairly straight, mostly following the lyrics and musical choices of the original but involving some changes in instrumentation and sound that reflect the abilities and interests of the musicians recording the cover. Others are more transformative, like Hendrix's cover of 'All Along the Watchtower' and Franklin's cover of 'Respect.'

It is common to suggest that covers exist on a continuum. For example, Mosser distinguishes different types of covers and writes that 'if we take "cover song" as a genus, we can identify a continuum of species that ranges from an attempt to reduplicate a given song or performance to a parody that maintains only the most minimal connection with the song being parodied' (2008). Deena Weinstein writes similarly, 'Covers exist along a continuum from those that are drastically different from, to those nearly identical to the original' (2010: 245).

This treats the types of covers as a matter of degree. That is, the difference is just how much the cover sounds like the original. However, a mimic cover need not and generally will not sound exactly like the original. A poorly executed mimic sounds quite different, and a fairly straight rendition can sound more like the original than a botched mimic does. What makes one a mimic and the other a rendition is their context of creation and appreciation. As I argue in the next chapter, the standards for what counts as a *good cover* are different for mimic covers than for rendition covers.

Moreover, thinking in terms of a continuum invites thinking of differences just along one dimension. This is plausible for a mimic cover, where any difference from the original is a demerit. For a rendition cover, however, differences can occur along many different dimensions. Changes to all the particular notes may preserve the harmonic progression. Conversely, a change in a few notes can make for a change from major to minor and thus change the whole mood. Changes to almost all the words in the lyrics might convey the same overall meaning, or a different change to just a few words could result in a substantial change to the meaning.

So placing covers on a continuum based on resemblance to the original fails to capture the distinction between mimic covers and rendition covers, and it smooths over important factors in the appreciation of rendition covers.

Conclusion

I have argued that understanding recent music better requires distinguishing songs, performances, and tracks. A cover is a performance or recording which covers an earlier recording. I have also argued that we should distinguish mimic covers from rendition covers— this distinction is put to more work in the next chapter.

Interlude: Cover Bands

In August 2014, Christy Mag Uidhir at the blog *Aesthetics for Birds* held a contest which asked philosophers to answer, in 50 words or less, the question 'Can a band be its own cover band?'

Figure 4: The announcement of contest results.

In a basic sense, a cover band is just a band that plays cover songs. Lots of cover bands play songs from many different sources, songs connected perhaps just by genre or sound. Other cover bands— so-called tribute bands— aim to reproduce the music of a specific band or artist. These often have humorous names, like Fan Halen (a Van Halen cover band), BC/DC (an AC/DC cover band from British Columbia), or Iron Maidens (an all-female Iron Maiden cover band).

The contest was prompted by a conversation I had with Christy, and he asked me to judge it. So I did not write an entry. However, I felt that the contest prompt suggested an obvious *No!* answer. I expected to see entries

https://doi.org/10.11647/OBP.0293.10

going something like this—

1. By definition: A cover version is performed by someone other than the original artist.

2. By definition: A cover band of X plays cover versions of X's songs.

3. The original band is the original artist and not someone else.

Therefore, the original band cannot be a cover band.

Whatever I might have written then, that is no longer something I would write. The dictionary definition of 'cover' supports the first premise of the argument, but that definition is subject to all of the problems that I discussed in Chapter 1. The word 'cover' is not precise enough to support such an argument. We could *make* it more precise by stipulating a definition, but that would be arbitrary and doctrinaire.

Another strategy— the one pursued in most of the contest entries— is to argue that a band can, in some sense, be a different band than they originally were. Here is Bree Morton's winning answer:

> Sure! Imagine a band that secretly re-unites, pretends to be a cover band of itself, and is taken as such by its new audience. Whether they fess up or not, by posing as a cover band successfully and being taken as one, they have become a cover band of themselves.

Jim Hamlyn earned honorable mention for composing a similar thought, less precisely but in verse:

> A band can be a cover band
> By a form of exaggeration and
> By mocking, shamming or otherwise hamming
> Their art as if by another's hand.

These arguments play on an ambiguity of what it is to be the same band. The band in the scenario is the same band in one sense because it has the same members playing the same parts. The band is a new band in another sense because band identity is a social fact, constituted by the members presenting themselves in a certain way and being accepted by the audience.

It is contentious whether one of these approaches to band identity is the right one. See, for example, the disagreement between James Bondarchuk and Ley Cray over whether the band Black Sabbath continued to be the same band after Ozzy Osbourne left (Irwin 2013: chs. 11–12).

3. Listening to Covers

In this chapter, I use the distinction between mimic covers and rendition covers to understand the evaluation and appreciation of covers. Before turning to that, we should talk about the fact that covers get a bad rap.*

The dim view of covers

People often take a dim view of covers. Simon Frith claims that a 'cover version is almost always heard as bad' (1996: 69). He recounts, 'One aspect of learning to be a rock fan in the 1960s was, in fact, learning to prefer originals to covers. And this was, as I recall, something that did have to be learned: nearly all the records I had bought in the late 1950s had been cover versions' (1996: 70).

This was well set in popular consciousness by 1973, when Ralph J. Gleason wrote in *Rolling Stone*, 'One of the most interesting developments of the past decade and the predominance of the performer/composer has been the relative absence of "cover" versions of hit songs.' As I discussed in Chapter 1, this is an exaggeration. There were still covers. Gleason acknowledges as much but insists that 'the drive for individuality in recent years has prevented most possible instances of this and it's something to be thankful for.' The change is not whether there were covers, then, but how they were viewed. What Gleason calls the 'predominance of the performer/composer' was actually the rise of a certain ideal— the model of the all-around genius who writes, performs, and records their own songs. This expectation that one person should write and perform the songs, along with conceptions of individuality and authenticity, make covers seem like a bad thing.

Standard 1960s exemplars of the ideal are Bob Dylan and the Beatles, but it requires a certain selective memory to think of them that way.

---

*Parts of this chapter are adapted from work I coauthored with Cristyn Magnus and Christy Mag Uidhir (Magnus et al. 2013).

 https://doi.org/10.11647/OBP.0293.03

Although Bob Dylan is the archetypal singer-songwriter, he is a strange source for the dim view of covers. He both played a lot of covers and wrote songs that were widely covered. His songs are lyrically powerful but lend themselves to musical reimagining. It is unlikely that they would have penetrated our culture so deeply if there had only been Dylan's own versions. Peter, Paul, and Mary had a hit with 'Blowin' in the Wind' just a month after the 1963 release of Dylan's version. The Byrds came to fame with their hit cover of 'Mr. Tambourine Man', and several other Dylan songs figure among their greatest hits. In the 1960s, as Tom Petty puts it, 'The Byrds was where you first heard Dylan' (Zollo 2012: ch. 1). Jimi Hendrix recorded the definitive version of Dylan's 'All Along the Watchtower.' Dylan's 'The Times They Are A-Changin'' was covered at least a dozen times within two years of its release and became an anthem of change, arguably a standard. I could easily extend this list, but the general point is that appreciating Dylan as a songwriter should give one a positive view of covers, of what they can do both for the writer of the song and for the musician recording them.

Once the Beatles gave up touring and dedicated themselves to studio production, they just made original tracks. So they provide some basis for the dim view of covers. Describing his own start in the 1960s, Petty comments that most bands emulated the line-up of The Beatles and 'everybody knew that they wrote their own songs' (Zollo 2012: ch. 1). In the time he is describing, Petty was playing in a cover band. And the Beatles also got their start playing covers. By one count they had recorded at least twenty-four covers of American rock and roll tracks by 1965 (Zak 2010: 236).

Theodore Gracyk gives several examples of other musicians who got started doing covers. He writes, 'Even as the [Sex] Pistols rose to national fame in the British media, cover versions dominated their live sets' (2001: 7). Regarding the Rolling Stones, 'their early career was dominated by performing and recording cover versions of blues and R&B songs they learned from records' (2001: 15).

This should be no surprise. There is not much alternative to starting out playing covers. Perhaps one could come into songwriting just at the same time one came into playing an instrument, playing one's own songs right from the start. More realistically, one might just play standards— songs that either predate recording or which do not have a canonical recording. Nevertheless, most musicians naturally start with covers. The pop singer Halsey comments, 'When I first started out, I was covering other people's music and doing that… helped me explore what my voice was… [to] pick up on the different things that other artists were doing that maybe I would do for myself or maybe I would have done differently' (Glamour 2018a). Hayley Kiyoko comments similarly, 'I feel like that's how you… start really writing your own music is adding beats and kind of incorporating your own twist to the songs that you love. And that's how I started, too.…' She adds, 'I… grew up playing guitar and singing songs that I wish I had written…' (Glamour 2018b).

This suggests a qualified version of the dim view according to which covers are a thing that a musician should only do when they are starting out. On this view, covers are training wheels to be removed once a band or artist has gotten the balance of serious songwriting.

Admittedly, original songs can provide better marketing. An artist or band is able to establish themselves— to define their brand— not only with their own sound and style but also with their own songs. There is a related financial incentive, because copyright does not protect sound and style. Suppose someone records a song, has enough success with it that others decide to cover it, and one of the covers becomes a hit. The recording artist on the earlier version adds to the song's cachet but does not get paid for the later hit, while the songwriter gets a royalty for every sale. As an example, consider Lori Lieberman, who was involved in writing 'Killing Me Softly' and recorded the original version— but she was not credited as a songwriter and so got no money when Roberta Flack and (later) the Fugees recorded hit versions. So being both the recording artist and the songwriter provides a kind of insurance. This leads to many recording artists being listed as cowriters on songs they record, regardless of their actual contribution. After Elvis Presley was an established star, his manager made it a regular practice that 'younger songwriters had to sign over half of their publishing rights to any song the King was going to record' (Padgett 2017: 162). Although giving out songwriting credit as a way of redistributing royalty money is something that still happens in pop music, young artists do not have the heft to demand it. So it is better for them to have their breakthrough song be one they wrote rather than one they covered.

However, there is also a financial incentive to keep playing covers. As Tony Kirschner notes, for small-time musicians, 'playing covers is far easier and more lucrative' than playing original songs (1998: 251). It is easier to get a gig playing a bar or an event as a cover band than as a band playing

Figure 5: A promotional photo of the King Beats, taken at Lowery Castle in
Middleville, New York circa 1965.
Image courtesy of Joe Chapadeau.

unknown original material.

This has been true for the whole history of rock and roll, with local music scenes supporting countless small bands. As one example, take the King Beats from Little Falls, New York. They played teen dances and night clubs, making enough money to pay their way through school. They learned to play songs from the radio and played what was popular. These were covers in the original sense of *coverage*. The King Beats had only one original song. Band member Joe Chapadeau wrote the bouncy, Dylan-inspired 'Nothing About Life At All' so that the band could appear on the regional TV program *Twist-a-rama USA*. Performances on *Twist-a-rama* are collected on a compilation album that has one song from each of several different bands, and Chapadeau comments that most of the bands had just the one original song.

Cover bands might have no chance of making it big, but bands who play original material probably will not make it big either. Very few bands make it big. Even the longshot at success may depend on being perceived as original, however, because of the sentiment that 'good rock music must be "original" and not reminiscent of past [work]' (Kirschner 1998: 253). This tension leads some musicians to use different names for cover projects than they use for original projects. The same ensemble might opt to use different band names— at once to have a cover band (so as to book gigs) and to avoid the suggestion that their original-material band is just a cover band (in hopes of making it big).

Although many artists develop original material in order to launch their careers, that does not mean they give up playing covers. Recalling the early days with his band Mudcrutch, Tom Petty says, 'We worked really hard at originals. Because we knew that there wasn't any way out unless we did. There wasn't any way to get a record contract if you were just covering the Stones' (Zollo 2012: ch. 1). Yet he did not stop playing covers. Commenting on the situation decades later with his band the Heartbreakers, he says, 'I always like the rehearsals better than the shows. Because we play *everything*. Though we might never play what we're going to play in the show, we play things that we all know, just to keep ourselves amused, and to get our chops up. So covers are a lot of fun' (Zollo 2012: ch. 12).

Singer-songwriter Shawn Colvin similarly started out playing covers. She says, 'When I was playing club gigs every night, I was always looking for new material— songs that I could do unexpected things with and interpret in ways that hadn't been thought of before.' Although she was signed to a major label in 1989 to record her original songs, she continued to work on covers. 'So as soon as I got signed,' she says, 'I let them know that I had this cover record that I've wanted to do, and I've been keeping notes on it ever since' (Cummings 1994: 14). That project was released as her third album, *Cover Girl*, in 1994.

The upshot of all this is that the dim view of covers is wrong, even in a qualified form. Despite the complex economics of it, there is nothing wrong with playing covers. There can be terrible covers, of course, but there can also be great ones.

Evaluating mimic covers

For a mimic cover, the aim is for the resulting sound to be exactly the sound of the original track. So the standard for a mimic cover is what Lydia Goehr calls the *ideal of perfect compliance* (Goehr 2007: 99).

This is the same ideal as for a sound system when it is used to play a track. For the sake of concreteness, imagine that you listen to the digital stream of a track played on a low-end gaming headset. If you complain about the tinniness of the sound or about the weak bass, then you are making technical rather than artistic judgements. If you recognize beauty or virtuosic vocals, then you are responding to the track itself. You might make both sorts of judgment at the same time, judging it to be beautiful despite the poor sound quality. In any case, though, you are assessing the sound playing through the headphones in relation to the track itself.

The same lesson holds for a mimic cover, since the ideal of fidelity is the same— that is, any evaluation of a mimic cover considers it in relation to the original. If you appreciate the interpretive choices in the new recording or performance, it reflects on the original track. The musician making the mimic is trying *not* to make any new interpretive choices, but instead just to follow the same interpretation as the original. If you appreciate the skill involved in a mimic cover, that is an assessment of the covering artist but in relation to the original. Their skill is directed at reproducing the sound of the original as closely as possible.

To make the point in a different way: Suppose you are inspecting a replica of a famous sculpture. Finding the replica to be beautiful is either a response to the same beautiful features as in the original (where the replica is accurate) or finding a demerit in the object as a replica (because it has features the original does not). Note that it is different than looking at an original painting of a beautiful subject. In that case, you are not just taking the painting as a mere echo, as documentation of what the subject looks like. An original painting can portray an ugly thing as beautiful. A mimic cover is importantly not a representation of the original, but rather a replica.

This highlights the fact that there is a limit to how great a mimic cover can be. Even if you admire the artisan who makes a replica, you admire merely their craft. The perfect replica would be interchangeable with the original— as good as the original, but no better. Contrawise, there is no limit to how bad a mimic can be. So the dim view of covers often depends on the unstated assumption that all covers are mimic covers. When Sandy Brown notes the distaste for covers, for example, he describes them as 'near copies of original recordings' (1968: 622).

We can see this by considering two covers recorded in the mid-1970s for *Top of the Pops*, a record series featuring mimic covers of recent hits. Tracks for the records were recorded quickly, with little time for rehearsal. Tony Rivers, who was one of the uncredited musicians, recounts the effort they put into their cover of Queen's 'Bohemian Rhapsody.' After spending the day in the studio recording other tracks, he spent hours listening to Queen's original and working out the vocal arrangement. After recording, they spent more time mixing and producing than usual. Given the complexity of the original, the mimic cover is impressive. According to Rivers, the cover even got some media attention— with one radio host playing both the original and the cover, cutting between the two (Rivers 2007). Rivers and the other musicians are clearly having fun with their performances, but the cover lacks some of the grace and vitality of the original. In short, it is a pretty good mimic cover because it approximates the original.

Contrast that case with the *Top of the Pops* cover of Stevie Wonder's 'Superstition.' Whereas the original has Wonder playing parts on clavinet and on Moog bass, the cover substitutes guitars for the clavinet and bass guitar for the synthesizer. The cover also lacks most of the rhythmic complexity of the original. The result is a disaster.

Evaluating rendition covers

Unlike mimic covers, rendition covers are not beholden to the ideal of perfect compliance. That means that we should not evaluate rendition covers by their sonic fidelity to canonical recordings. So how should we evaluate them?

One could say: Whereas we *compare* a mimic cover to the canonical track, we *contrast* a rendition cover with it. A mimic cover fails insofar as it departs from the canonical version, but a rendition cover is successful only insofar as it departs from the canonical version in artistically interesting or rewarding ways. Michael Rings suggests something like this when he writes that covers invite 'contrastive appreciation', that is 'the appreciation of the cover insofar as it musically contrasts with the previous version' (2013: 59).

To underscore the difference from mimic covers and classical practice, let's call this proposal the ideal of *rewarding deviation*. The ideal of perfect compliance sets a uniform standard for a mimic cover, because it can be satisfied only by sounding one specific way. The ideal of rewarding deviation is more complicated. It might be satisfied in different ways, because there is no limit to the ways a rendition cover might differ from the original.

Although I said things like this in some earlier work, I now think this view would be a mistake.

First, the rewarding features of a rendition cover are not just the ways in which it diverges from the original. It may also be rewarding for the ways in which it matches the original. For a mimic, which aims at matching the original in all respects, the differences are all failures of skill. For a rendition, however, the differences reflect interpretive choices— that means that the similarities, too, reflect interpretive choices. This is especially clear for lyrics, because opting to sing even the same words can mean something different because the context of the cover is different than the context of the original. I explore this point further in the next section.

Second, a rendition cover can be a great version of the song apart from any consideration of the original. It can be beautiful, powerful, or moving— features that it just has on its own without considering its relation to the original. This point will come out further on.

Lyrical changes

A rendition cover may have different lyrics than the original while still being the same song. Take the example of Willie Nelson's 1993 cover of Paul Simon's 1986 'Graceland.' In his original version, Simon sings, 'There is a girl in New York City / Who calls herself the human trampoline.' Willie Nelson sings 'a girl in Austin, Texas' instead.

How should we think about this difference?

Gaynor Jones and Jay Rahn suggest that the potential for a cover to be different than the original means that the covering artist simply does not care what the original artist had in mind. They write, 'This range of variability is probably related... to the audience's and performer's lack of concern about the composer's intentions' (1977: 85). This seems wrong, at least in this case, because Nelson's choice of 'Austin, Texas' is arguably in line with Simon's intention.

Gracyk points out that songwriters typically have 'open-ended intentions' such that 'elements of a lyric that seem to refer to concrete things and situations are merely exemplary.' As such, 'specific places, objects, and people... function as placeholders for ideas more than as references to the individuals they mention' (2001: 66). Simon's song is not about New York especially. The verse continues that our lives are 'tumbling in turmoil', and that message is the verse's more central function. To use Gracyk's word, New York City is serving as a *placeholder* for where the girl is from. Austin, Texas fits Nelson's idiom better, and it does not have any significance which would undermine the central function of the verse. So making that substitution respects Simon's intention— not for what specific words to sing, but for what open-ended meaning the lines are supposed to have.

This view is problematic if intentions are private, mental things. Consider a different example: Tom Petty and the Heartbreakers' song 'Listen to Her Heart' opens with the lines, 'You think you're gonna take her away / With your money and your cocaine.' Suppose one were to cover it and replace 'cocaine' with 'champagne.' Both are things that might tempt someone. One might suppose that Petty has an open-ended intention which allows for such a substitution. As it happens, Petty's intention was more specific. The record company wanted him to change 'cocaine' to 'champagne', but he refused. Petty offers this reason: 'Because it would have made it a different song. I didn't really see the character as caring about the price of a bottle of champagne. Cocaine was much more expensive' (2012: ch. 16). I happen to know about this because Petty discussed it in an interview, but what if he had not? If his record company had not asked for the change— giving him the occasion to refuse— then there would be no way for anyone else to know that his intention was specific rather than open-ended.

Petty says that changing that one word would make it 'a different song.' Even if that seems too extreme, it is possible for a rendition to change so much that it is no longer the same song as the original. That is a matter of metaphysics rather than appreciation, though, so let's set it aside for now— see Chapter 5.

I have to admit that I have no idea what Simon's intention was when writing 'Graceland.' Perhaps, because of his personal experience, he had in mind a character who was specifically from New York City. Perhaps the girl in New York City was a specific person he knew, and her inclusion in the song was a shout out to her. Nevertheless, I think that Nelson's word substitution when singing 'Graceland' is fine *regardless* of what exactly Simon might have been thinking. That is, I do not think that the songwriter's intention carries all the weight. Rather, word substitutions have to fit within the overall meaning of the song. The meaning of the song depends not just on intentions, but also on the content of the song and the musical conventions of the community.

Note that changes in meaning are not simply a matter of changing words. As Dai Griffiths comments, a 'rendition can be "straight" but context changes meaning' (2002: 61). He offers an extended discussion of women covering Bob Dylan's 1966 'Just Like a Woman.' In her 1970 cover, Roberta Flack changes the viewpoint of the song to be first-person. For example, where Dylan sings 'And she aches just like a woman / But she breaks just like a little girl', Flack sings 'I ache just like a woman / And I break up like a little girl.' Dylan dismisses Flack's version, saying in a 1975 interview that 'she got the words wrong' (Travers 1975). However, that seems to miss the point. If Flack had sung exactly the same lyrics, it still would have offered different interpretive possibilities than when Dylan sang them. Griffiths also discusses Judy Collins' 1993 cover. The instrumentation and vocal style are changed, but she follows Dylan's lyrics verbatim. Griffiths suggests several interpretive possibilities: Collins can be heard as addressing another woman as 'a friend or even a mother to a daughter.' Although she might be addressing a lover, Griffiths recounts that when he taught the example his students would go to great lengths to avoid interpreting it that way. He suggests that 'we bring slightly different expectations to female covers that allow a lesbian interpretation, where with male covers we might leap more swiftly to a gay interpretation.' Regardless, there is another interpretation which neatly avoids the question of who Collins is addressing. Her version can be seen as 'a strict rendition of the original in the manner of classical music' (2002: 53). There may be no deeper significance to Collins singing the same words as Dylan than that Collins approaches covers in that way.

To take another example: Joni Mitchell's 'Big Yellow Taxi' is a meditation about progress and loss, until the last verse. Then it becomes personal with the lines 'Late last night, I heard the screen door slam / And a big yellow taxi took away my old man.' Note that the phrase 'my old man' could either mean her boyfriend or her dad, but the context of Mitchell's other work makes it clear she means her boyfriend. Harry Styles does a cover and sings the same words. In his version, I find that I hear the phrase as father rather than as boyfriend. It is a subtle difference, but the change results even from Styles singing the *same* words. Other cover artists opt for different words. Counting Crows substitute 'took my girl away', replacing Mitchell's almost-rhyme with a line that does nothing like rhyming. Keb' Mo' changes the lyrics to 'Late last night, she heard the screen door slam / And a big yellow taxi took away her old man.' This puts it at a distance, so the singer is not singing about his own loss but instead about hers (Mitchell's?). Bob Dylan changes the lines to 'Late last night, I heard my screen door slam / A big yellow bulldozer took away the house and land.' Dylan's cover of 'Big Yellow Taxi' thus leaves out the only reference in the song to the titular taxi. It might be seen as conveying the same core meaning, though, since he retains most of the lyrics and the central themes of progress and loss.

These examples underscore why it is shortsighted to evaluate rendition covers just for the ways they depart from the original. Even singing the same lyrics reflects a choice to do so, one which can be considered in evaluating the cover.

Genre shifts

Rendition covers are often in a different genre than the canonical recordings. Consider the Cardigans' covers of 'Iron Man' and 'Sabbath Bloody Sabbath', both originally by the heavy metal band Black Sabbath. Also worth a listen is their a cappella cover of Ozzy Osbourne's 'Mr. Crowley.' It is hard to say exactly what genre the Cardigans are. I have seen suggestions such as alternative, indie, ambient pop, and Swedish pop/rock— I am not sure that the last of those is really a genre. Regardless, they and their covers are certainly *not* heavy metal. Lead singer Nina Persson has said that their covers of Black Sabbath 'take their rock songs and make them really wimpy pop music.'

The Cardigan's 'Iron Man' tells the same story using the same words as Black Sabbath's original. Yet one could hardly mistake one version for the other. The emotional content is different. One might find Black Sabbath's faster, louder version to be more threatening and the Cardigan's softer cover to be more melancholy. This difference in emotional valence is not enough to make it a different song, but it certainly makes for a different listening experience.

Michael Rings argues that covers which are in a different genre than the original are interesting for two reasons.

In the first place, Rings writes that 'the interest generated by genre resetting in rock covers is produced via a manipulation of listener expectations' (2013: 57). His idea is that these covers can surprise us by treating familiar material in an unfamiliar way. 'My proposal,' Rings writes, 'is that much of the pleasure in listening to these covers comes from following a familiar song's progress through an unfamiliar (relative to the song) stylistic landscape: *how will this next passage sound within the conventions of this new genre?'* (2013: 60) Of course, although one can attend to differences in how the song sounds in its new setting, it does not follow that this is the predominant interest of such covers. First, one can only sincerely *wonder* how the rendition goes on the first listening. Hearing it again, one *remembers* how it went. If anticipation and surprise were the primary appeal of a genre-shifted cover, then it would stop being interesting after one had heard it enough to be familiar with it. They would be disposable novelties. Many genre-shifted covers, like the Cardigans' 'Iron Man', are not like that. Second, there is a way in which rendition covers can be aesthetically interesting when there is no change in genre. For example, all the covers of 'Big Yellow Taxi' discussed in the previous section are in roughly the same genre. Surely at least some genre-shifted renditions are rewarding in *that* way, rather than in virtue of the genre shift itself.

In the second place, Rings maintains that 'genre resetting in covers can also generate a more hermeneutic brand of interest by providing an intertextual dimension that may serve to enrich a listener's interpretative engagement with the song' (2013: 57). Elaborating on this dimension, Rings suggests that such covers are 'a way for rock artists and appreciators to engage in a critical dialogue about rock itself— its history, its categories, its multifaceted culture' (2013: 63). His idea is that changing the genre of a song raises questions about how to categorize and think about the music, and that certainly can happen. When the band Social Distortion recorded a punk cover of Johnny Cash's 'Ring of Fire', they were responding to discussions of 'what's punk and what's not punk' and legitimizing the subgenre of cowpunk (Hodge 2017). However, I think it overintellectualizes the process to say, as Rings does, that 'contemporary rock audiences... find interest and entertainment in sustaining such a conversation' (2013: 63). Social Distortion's cover is at most implicitly an argument, and one can legitimately appreciate it without making the historical and social dimensions explicit. For someone who just likes classic country and punk, the enjoyment of cowpunk need not be an occasion for conversation.

Some genre-changing covers are successful just because the resulting version sounds good. The Cardigan's 'Iron Man' is a beautiful and haunting track. It has these features independently of whether or not it shares them with the original. A listener can respond to these features without having the original in mind.

Two modes of appreciation

What these examples show is that there are two broad approaches to appreciating a rendition cover.

First: Appreciation without having the earlier track in mind. Listen to it just as you would to an original recording of that song. We might call this an *aesthetic* mode of evaluation, appreciation informed by *immediate features* of the version.

Second: Appreciation in relation to the canonical track. Consider both how the cover diverges from its source and how it remains faithful. Employing this approach, we might pay attention to musical features (sounds), to meaning (significance), or to both. In some cases, keeping things the same in the cover will have a different significance than in the original. In others, considerable musical changes result in little change to the meaning. We might call this an *etiological* mode of evaluation, appreciation informed by *provenance*.

I do not want to overburden my account with extra jargon, and the distinction can be put in plain language: *There are two modes in which one might evaluate a rendition cover— on its own or in relation to the original. Neither of these modes is necessarily better than the other.*

Discussing how they decide which songs to cover, Paul Dempsey of the band Something For Kate says that it has 'got to be something that you feel like you can sing with as much honesty and conviction as the original, as the person who wrote it. Something you can identify with and kind of sing honestly' (JJJ 2021b). Bands who take this attitude try to produce renditions which at once capture what they like about the original while also sounding like it is theirs. In addition to examples discussed above, take R.E.M.'s cover of Leonard Cohen's 'First We Take Manhattan.' The track was recorded for the 1991 tribute album *I'm Your Fan*, a project explicitly aimed at making a younger audience more familiar with Cohen's work. Yet, as Ray Padgett notes, R.E.M.'s version of the song makes it 'sound like it was theirs to begin with' (2020: 62). It is rewarding to listen to it both as an R.E.M. track (without regard to it being a cover) and as a variation on Cohen's less driving original. There is value in considering it in both modes.

Originality and rocking out

One may object that it is impossible to judge the originality of a rendition cover without comparing it to the earlier version— so, the argument would conclude, one can only properly evaluate a cover by considering it in relation to the canonical version. For example, you might be less impressed with Joan Jett and the Blackhearts' 1981 'I Love Rock and Roll' when you learn that it is a cover of a 1975 version by the Arrows.

As a first answer to this objection, note that originality is only one artistic virtue. The 1981 cover *rocks*, which one can notice without having the original in mind. The comparative judgement that the cover *rocks harder* than the original— which it does— is different than the non-comparative judgement that it rocks. Why should we suppose that the relative properties of originality and rocking-harder are more important than the immediate property of rocking?

Pressing the objection, one might insist that proper appreciation of a cover still requires considering originality as one factor among others. If that were so, though, how many possible influences would one need to take into account?

Consider 'The Last Train to Clarksville', which was written for the Monkees by Bobby Hart and Tommy Boyce. It was meant to sound like a Beatles song and directly inspired by the Beatles' 'Paperback Writer'. Although it is not a cover, it is not very original when compared to its influences. A cover might be unoriginal in the same way, by drawing too much from other influences.

Tom McDonald complains about formulaic punk and heavy metal covers, writing that "within six seconds you know what the next three minutes holds for you" (2021). This predictability is not internal to the cover itself, but instead because the covers follow a standard formula. You know what to expect because you have heard other covers that use the same formula. So the judgement that such a cover is unoriginal does not depend on comparing it to the canonical version of the song, but to other tracks and covers in the same genre.

Returning to the example of 'I Love Rock and Roll': If we consider just the Arrows' earlier version, then the 1981 cover may seem to add little. However, Jett made an earlier attempt in 1979, which she recorded with Paul Cook and Steve Jones of the Sex Pistols. The 1979 cover is uncompelling. It does not rock the way that the 1981 cover does. So what is added in the 1981 version, although subtle, is nothing trivial or easy. If it were, they would have added it in the earlier version.

So it seems to me that all or most of the value in Joan Jett and the Blackhearts' version of 'I Love Rock and Roll' is found in its immediate features— that is, in rocking out to it. There is little appreciative gain in listening to it along with the original. Even if you disagree, surely there *could* be such covers— ones which stand on their own as works of music.

One could try to turn this around to mount a different objection to the two-mode account. If the 1981 cover of 'I Love Rock and Roll' is best appreciated for its immediate features, rather than in relation to the original, then one of the modes is decisively better than the other. This objection misunderstands the point of distinguishing the two modes of appreciation. The point is not that both modes are equally rewarding or that both will always offer significant rewards. Rather, there is no way to know *a priori* how rewarding either mode will be. In some cases, both approaches will be rewarding. In other cases, one approach will dominate. In still other cases, a cover will be terrible and neither approach will be rewarding. In order to figure out how a specific rendition cover fares, we have to listen. General principles will not tell us.

Is a cover better if it rewards appreciation in both modes than if it rewards appreciation in just one? Yes, all things being equal— but all things are never equal. The immediate experience of listening to Joan Jett and the Blackhearts' 'I Love Rock and Roll' would be very different if it were also wildly different than the Arrows' original. So I am not saying that covers ought to be appreciable in both modes or that covers which reward appreciation in both modes are better. That would underwrite the conclusion, just on principle, that R.E.M.'s 'First We Take Manhattan' is better than Joan Jett and the Blackhearts' 'I Love Rock and Roll.' I like both and refuse to decide between them. Even if you think them comparable and consider one to be better than the other, it must be for particular features of them. The details matter.

When artists try to hijack their own hits

Although it is not usually considered a *cover* when a musician records a new version of one of their earlier tracks, the framework for appreciating covers can be helpful in thinking about how to assess those remakes.

In 2012, the band Def Leppard recorded new versions of some of their classic songs. The recordings were part of a struggle with their label, which had control over the original versions of the band's songs. They called these new versions 'forgeries' and tried to make them sound as much as possible like the originals. Lead singer Joe Elliot said of the effort, 'I had to sing myself into a certain throat shape to be able to sing that way again' (Rolling

Stone 2012). These new versions are not literally forgeries, of course. And whether we call them covers or not, these new tracks are mimic versions. The commercial rationale is exactly the same as for predatory covers which are supposed to take sales and radio plays which might have gone to the original. The artist doing the mimic in this case is the same as the artist who recorded the original, but a similar effort is required because it is their older selves trying to sound just like their younger selves did. Evaluation of these new tracks should work the way the evaluation of a mimic cover does. Success or failure is measured by fidelity to the original.

At about the same time, Suzanne Vega released *Close-Up*, a four-album series featuring acoustic versions of her earlier hits. Part of her motivation was that she did not own the masters to her earlier hits. Unlike Def Leppard's forgeries, however, Vega's new versions are not meant to sound like the tracks from the earlier albums. Instead, they are intended to capture more of the sound of her live performances. Earlier, many artists produced recordings in that style for MTV's *Unplugged*. The difference here is that Vega's *Close-Up* versions are explicitly meant to be played instead of her original tracks. Although it may seem wrong to call them covers, they are renditions. Evaluation of them follows the same lines as the evaluation of rendition covers.

After a bitter dispute with a music manager who had bought the masters to her first six albums, Taylor Swift ultimately decided to rerecord them. Since she hopes that her new versions (the ones she owns) will displace the originals in digital streaming and licensing, she has made the instrumentals and production of the new versions sound as much like the originals as possible. The first of these do-over albums was released in April, 2021. *Fearless (Taylor's Version)*, a remake of 2008's *Fearless*, debuted at #1. The album includes six bonus tracks, songs written at the time of the earlier album but recorded in 'a hybrid style that split the difference between the two eras' (Willman 2021). The core of the new album is mimic versions— whether we call them covers or not, the ideal is perfect compliance. The bonus tracks are just originals— songs written earlier in her career, but without any earlier recording to compare them to.

Brandon Polite argues that we should understand Taylor's Version differently. He suggests that her project is an act of aesthetic disobedience, expressing themes of independence and empowerment (Polite, forthcoming). If that is right, then listening to the tracks as mimics misses the point.

Instead, the entire album should be understood as props in a bigger performance.

Regardless, Def Leppard's 'forgeries' and Vega's *Close-Up* versions are not part of such a grand drama. Rather, they reflect what Brown calls the 'jackal thinking behind cover versions' (1968: 622). That is, they are intended to earn royalty money which would otherwise have gone to the original recordings.

Is the distinction grounded in intentions or in appreciative standards?

When I introduced the distinction between mimic covers and rendition covers in the last chapter, I did so in terms of the intention with which they are made: Mimic covers are intended to sound like the original, while rendition covers are intended to sound different. In this chapter, I have argued that the appreciative standards for the two are different: Mimic covers are subject to the ideal of perfect compliance, while rendition covers are not. It is not clear to me which of these— intention or appreciative standards— should be taken as the defining feature that separates mimic covers from rendition covers.

The two sets of criteria usually yield the same result, but perhaps it is possible for them to diverge. Consider the *Top of the Pops* cover of 'Superstition', discussed earlier. The musicians had the task of making a track that sounded as much like the original as possible, given a limited amount of studio time. With an eye to intention, it is a mimic cover— and the result is pretty bad by the standard of mimic covers, hampered at the outset by the fact that guitars are playing keyboard parts. The cover might seem less terrible treated as a rendition cover. For a rendition, the substitution of one instrument for another is an artistic choice. For example, Jeff Beck opts for guitars in his cover of 'Superstition' (released with the supergroup Beck, Bogert & Appice a month after the *Top of the Pops* cover). No surprise, since Beck is a guitarist. However, Beck's version also shifts the sound from funk to rock with corresponding musical changes throughout. His intention clearly makes it a rendition cover, and it is well regarded. Considered as a rendition, the *Top of the Pops* cover follows the original too closely— so it is still pretty bad. The point, though, is that it is differently bad if considered as a rendition cover.

It is possible to imagine a case which is intended to sound the same as the original but fails to do so in a way that is unintentionally novel and interesting. Considered as a mimic cover, it is terrible. Considered as a rendition cover, it is great. In such a case, the intention would make it a mimic cover. It would be more rewarding to consider it as a rendition cover instead, so the pursuit of such rewards would count it that way. Take that as a thought experiment, because I do not have a convincing example.

If the difference between a mimic cover and a rendition cover were an intrinsic metaphysical fact, then the right answer would be the one that gets that fact right. However, the difference is not intrinsic. If it is determined by the intention, then it depends on a relation to the artist and context of creation. If it is determined instead by which categorization makes engaging with the version more rewarding, then it depends on the audience and the context of appreciation. So the difference between the two approaches is not a disagreement about what exists but instead about which context we want to emphasize. If we proceed from the point of view of the artist, then intention is primary. If we proceed from the point of view of the audience, then appreciation is primary. We can usually start from either end, because they yield the same outcome in most cases.

Considering whether a musician's recording should even count as a cover, Gabriel Solis writes, 'what matters is both how she plays these songs and how she and her audiences think about them' (2010: 311). These are the same considerations that make it the kind of cover it is. At least in principle, though, how a musician plays and how audiences think about it could diverge.

Conclusion

To sum up: It is inappropriate to take a dim view of all covers. How to sort good covers from bad ones depends on the kind of cover in question. A mimic cover, subject to the ideal of perfect compliance, is good just insofar as it echoes the canonical version. A rendition cover is more complicated. It can be appreciated for how it sounds. It can also be appreciated in relation to the canonical version— both for how it follows and how it diverges from the canonical recording, both for differences in sound and differences in meaning.

4. The Semiotic Angle

In the previous chapter, I proposed that a rendition cover can be appreciated in two modes: on its own (without consideration of the earlier recording that it covers) or in relation to the recording that it covers. There are a number of scholarly accounts of covers which put pressure on this view. Some scholars have argued that a cover *alludes to* or *pictures* the track that it covers. If such a relation were crucial to understanding the cover, then it would turn out that the cover could only be appreciated in relation to the earlier recording. The challenge can be met, but answering it reveals some interesting complexities.*

Reference, allusion, and hearing in

A number of philosophers have taken what Lee B. Brown calls a *semiotic angle* on covers, according to which 'a cover, unlike other recorded remakes of earlier recordings, stands in a referential relationship to the recording it covers' (2014: 193). Brown recognizes, as I did in Chapter 1, that this will not do as a definition of what it is to be a cover. Nevertheless, some covers are like that.

Referential covers

Some covers make explicit reference to the recordings that they are covering.

Consider the Meatmen's 1996 cover of the Smiths' 1984 track 'How Soon Is Now?' which switches the genre from alternative rock to punk and makes a subtle change to the lyrics. Morrissey, lead singer of the Smiths, sings each chorus as 'I am human, and I need to be loved.' Tesco Vee of the Meatmen echoes this in the first chorus, but changes it to 'I am inhuman and I need to be fucked' and '...I need to be killed' in the second and third choruses. The

---

*Parts of this chapter are adapted from work I coauthored with Cristyn Magnus, Christy Mag Uidhir, and Ron McClamrock (Magnus et al. 2013, 2022).

https://doi.org/10.11647/OBP.0293.04

new words do not serve the central function of the original lyrics. 'How Soon Is Now?' is a plaintive cry for the oppressively shy, and being loved, fucked, or killed are radically different prescriptions for such a condition. So the Meatmen are changing the meaning of the song. Moreover, the Smiths are (described uncharitably) a mopey band, and Morrissey (by his own declaration) celibate. Whereas we are invited to understand the 'I' in the Smiths' version of the song as Morrissey or someone like him, the tone of the Meatmen's cover does not invite us to understand it as being about Tesco Vee. Rather, the 'I' in the Meatmen's cover is most readily understood mockingly to be Morrissey. It is the mopey and asexual Morrissey whom Vee is saying needs to be fucked or killed. (This interpretation is supported by other work by the Meatmen. They have an original song titled 'Morrissey Must Die.')

Consider also the Screamers' 1978 version of Sonny and Cher's 1966 'The Beat Goes On.' The cover is harsher and more confrontational than the original, with considerably refigured music and lyrics. Sonny and Cher sing that 'Charleston was once the rage,' but the Screamers sing 'Anarchy's the current rage.' Where Sonny and Cher sing that 'Grandmas sit in chairs and reminisce,' the Screamers sing, 'Pop stars sit in chairs and reminisce. Kids today are right to make a fist!' After a line about cars being faster, the Screamers add that 'Sounds are moving faster, faster!' So the Screamers' version condemns slow, schmaltzy musicians like Sonny and Cher. It is both a cover of Sonny and Cher and a commentary on them.

A much-discussed example is Sid Vicious' 1978 cover of 'My Way' (Mosser 2008, Rings 2013). The song was written for Frank Sinatra by Paul Anka in the late 1960s, and Sinatra's version is canonical. In the cover, as Leonard Cohen quips, 'the certainty, the self-congratulation, the daily heroism of Sinatra's version is completely exploded by this desperate, mad, humorous voice' (Snow 1988). In the final verse of the song, Sinatra claims to be a man true to himself who can 'say the words he truly feels'— but Vicious inquires after a 'prat' who 'wears hats' and 'cannot say the things he truly feels.' So it is plausible to think that the considerably changed lyrics in Vicious' cover are in part a comment on Sinatra in the same way that the Meatmen's 'How Soon Is Now?' is a comment on Morrissey and the Screamers' 'The Beat Goes On' is a comment on Sonny and Cher. Responding to this suggestion, Nadav Appel argues instead that 'most of the changes are *non sequiturs* and mainly give the impression that Vicious forgot some of the original lines and replaced them with swear words on the spot' (2018: 448–449). Although Appel suggests that this question should be left to 'punk historians', the lines are not just drug-addled expletives. The words mean something, and they are plausibly about Sinatra. (That is not to say that my interpretation is necessarily right. As of this writing, the Wikipedia entry for 'My Way' indicates that the 'reference to a "prat who wears hats" was an in-joke directed towards Vicious's friend and Sex Pistols bandmate Johnny Rotten, who was fond of wearing different kinds of hats he would pick up at rummage sales.' This sentence is labeled *citation needed*, but perhaps it is true.)

In these punk covers of non-punk originals, the cover serves as an opportunity for the artist to thumb their nose at the earlier version. Doyle Greene (2014) refers to covers like these as *anti-covers*— 'anti' in the sense of *against*, because an anti-cover stands against the canonical version. Deena Weinstein sees punk as representing a 'new episteme' in which covers 'subverted the originals by transposing them to an irreverent musical attitude' (1998: 144). This is to overstate the matter, however. Even in punk, many covers are not anti-covers. First, some punk covers repurpose the original without providing commentary on the original. For example, the Dead Kennedys cover of 'I Fought the Law' changes 'the law won' to 'I won.' With other revisions to the lyrics, it becomes a pointed social commentary; as Greene writes, it is 'a protest song about the reduced manslaughter conviction Dan White received after he assassinated San Francisco mayor George Moscone and city supervisor Harvey Milk...' (2014: 42). Although it is somewhat obscured by the fact that Jello Biafra of the Dead Kennedys sings in the first-person, in the character of Dan White, the victory of the narrator over the law is meant to be understood as an absurd injustice. It is a cover of the Crickets' 1960 original or perhaps of the Bobby Fuller Four's better known 1964 cover, but it is not a rebuke of the Crickets or Bobby Fuller. Instead, the song is repurposed to provide commentary about then-recent events. Second, many punk covers are played straight. The Clash's 1979 cover of 'I Fought the Law' changes the musical style without any change to the story of the song. Shane MacGowan's 1996 cover of 'My Way', like many, reflects the Vicious cover musically but uses the lyrics from the Sinatra version. So, although some punk covers refer to the canonical version that they are covering, many others do not.

Next, consider an example from outside of punk: Ella Fitzgerald sings in her version of 'Mack the Knife' that 'Bobby Darin and Louis Armstrong.

They made a record (ooo, what a record) of this song. And now Ella, Ella and her fellas, we're making a wreck (what a wreck, such a wreck) of this same old song.' (These are the lyrics from a 1963 performance in Stockholm but are similar to ones she had improvised during a 1960 performance in Berlin.) One might object that Fitzgerald's 'Mack the Knife' does not count as a cover because it is a work of jazz and because covering is a category that structures rock and pop music. This sentiment is expressed by Michael Rings, who writes, 'A given jazz performance of a "standard" such as "Mack the Knife" certainly offers a new take on a song that has been both performed and recorded previously by countless other jazz musicians, but as a rule jazz-acculturated listeners will neither perceive nor appreciate it as a remake of any other specific performance or recording' (2013: 56). Although Rings mentions 'Mack the Knife', he does not mention Fitzgerald's version. Indeed, her version seems to undermine his claim, because she explicitly identifies it as a remake and refers to earlier versions. She even scats an impersonation of Louis Armstrong. Moreover, the versioning practices of pop music may be appropriate because Fitzgerald's version is at once jazz and pop. Bobby Darin's 1959 version was a crossover hit, reaching #6 on the *Billboard* R&B chart and #1 on the Hot 100 pop chart. Unsurprisingly, Fitzgerald's version is commonly described as a cover.

It is also possible for reference to occur in the musical elements. Albin Zak gives an example: In their recording of the Buddy Holly song 'Words of Love', the Beatles not only reproduce the timbre and tone of Holly's original but also add handclaps that echo the sound of another Buddy Holly track. Zak writes that 'the pat-a-cake effect of the eighth-note handclaps is a reference to another Holly track, "Everyday".' He continues, 'So while the Beatles track is at one level a cover version of a Buddy Holly song, it is also a more extended allusion to Buddy Holly's sound' (2001: 27).

In earlier work, my collaborators and I called covers like these *referential covers* and considered them as a separate category from rendition covers (Magnus et al. 2013). I now think of rendition covers more broadly, so that referential covers are a particularly interesting kind of rendition cover.

Saturated allusions

Other scholars have suggested that a cover can refer to the earlier version without any specific lyrics or sounds providing the reference. Theodore Gracyk (2007, 2012/3) considers cases where the musician or band connects to an audience, most of whom will be familiar with the original, with the intention of having their cover be understood as a reference and reply to the original.

Gracyk suggests that this is the case with tribute albums, which consist of new covers of songs where all the canonical versions are by the same artist. Considering two tracks on an Elvis Presley tribute album, he writes, 'When Bruce Springsteen covers "Viva Las Vegas" and the Jesus and Mary Chain covers "Guitar Man", each renders the song in their own style. But the musicians intend that we compare what they have done with the "originals", the Elvis Presley versions, and we are invited to interpret each new performance in light of [those]' (2007: 82). In these versions, there is not a particular lyric or riff that refers to the canonical Elvis track in the way that the eighth-note handclaps in the Beatles track refer to Buddy Holly's 'Everyday.' Instead, the entirety of the cover refers to the earlier track. Gracyk calls this *saturated allusion*. 'Normally, an allusion is a brief or relatively small aspect of a text,' he writes, but in these cases, 'Every aspect of the performance is to be treated as referencing all aspects of the earlier recording at parallel points in the performance' (2012/3: 41). The idea of *saturation*, in saturated allusion, is that the reference fills up the whole version rather than occurring in some specific lyric or sound.

Rings suggests, 'Over time, covers in general have become more allusive in regard to the objects of their remaking' (2013: 57). I am not sure this is true. Gracyk achieves this result by defining 'cover' to mean a version that makes a saturated allusion to an earlier recording— see Chapter 1 for reasons not to do this. If we consider versions that are typically called covers, then some of them are saturated allusions and others are not. Both kinds continue to be made.

The fact that some covers are made with the intention that the audience think about the original recording is enough, though, to make what Gracyk and others say about them worth some attention. Andrew Kania writes that 'it is impossible to properly appreciate an allusion without considering what it is an allusion *to*' (2020: 239). This suggests that the *only* possible mode of evaluation for a saturated allusion cover is to consider it in relation to the original. This threatens to collapse the two-mode view of appreciating rendition covers.

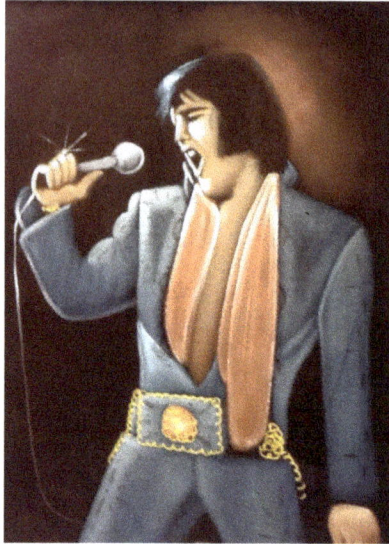

Figure 6: A vintage painting of Elvis Presley on black velvet. Although it is easy enough to see the details on the belt as just brushstrokes, it is hard not to see Elvis in the top part of the painting.
Photo by Mike Mozart, Creative Commons Attribution 2.0 license.

Hearing in

Jason Leddington (2021) points out that when someone listens to a cover, they might hear the original in it. This is more than just recognizing that it is a cover and thinking of the original. Rather, it involves experiencing the cover differently because of the original. As Deena Weinstein writes, 'In appreciating covers, one feels and recognizes similarities and differences between the original and the cover: the first plays in your mind's ear, while the other comes in through your ears from some playback device' (2010: 246). Albin Zak discusses a case which illustrates the possibility. He writes, 'However many cover versions I may hear of "Be My Baby," I can never separate what the song means to me from the image I hold in memory of Ronnie Spector's voice and Phil Spector's lavish production. Somehow, the cover performance resonates with the memory, and though the sound is all different, the meaning imparted by the original recording still comes through' (2001: 31–32).

This phenomenon of *hearing in* can be understood by analogy with the familiar phenomenon of *seeing in*, sometimes called *seeing as*. When I look

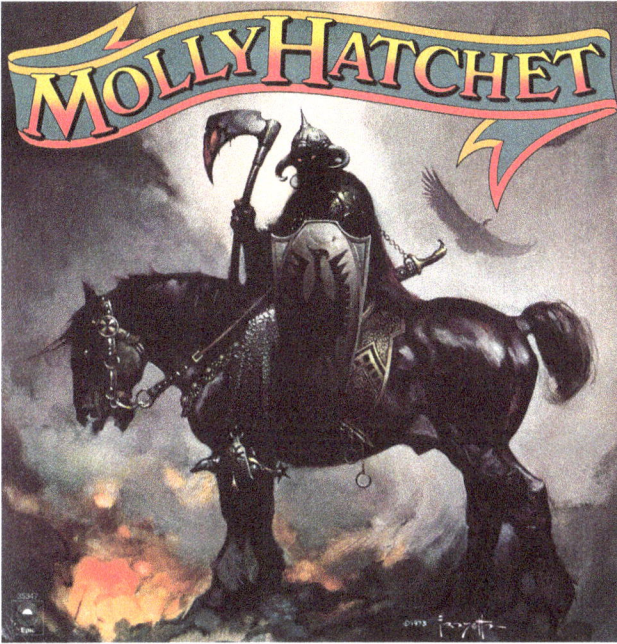

Figure 7: The cover of Molly Hatchet's eponymous debut album. The painting is 'The Death Dealer', by Frank Frazetta. Maybe you can focus your eyes past it so as to see just colors and shapes, but it is more readily seen as a dude with an axe on the back of a horse. That is *seeing in*.
Image used with permission of Frank Frazetta Jr. and the Frazetta Museum.

at a portrait, I typically experience it as seeing the subject of the portrait. For example, if I look at a painting of Elvis on black velvet, I see it as Elvis. (See Figure 6.) I do not first see the paint and then infer from its configuration that it is intended to be a picture of Elvis. Although it is possible to focus on the individual brushstrokes and the texture of the velvet, that is neither what I tend to do nor what I am meant to do when looking at such a painting. When I look at the painting in the usual way, Elvis is the object of my perception. Of course, I do not literally see the actual person Elvis Presley. Rather, I see Elvis in the velvet painting in the same way that I see the fictional Death Dealer in the cover of Molly Hatchet's debut album— as the phenomenological, intentional object of my perception, which need not correspond to any actual object out in the world. (See Figure 7.)

Likewise for sound: *Hearing in* happens when I hear some object or event as an intentional object presented through a more immediate presentation. For example, I might hear a friend speaking in the murmur of voices at a party. Leddington argues that in such a case I do not simply hear the *sound* of my friend. Instead, I hear my friend. Similarly, I hear the clavinet in the full sound of Stevie Wonder's 'Superstition.' Again, I do not simply hear the *sound* of the clavinet. Anyone who hears the track hears that, even if they lack enough understanding to distinguish the clavinet from the other instruments. Someone who cannot resolve the clavinet as a separate object of perception is hearing the sound but is not hearing the clavinet in the sense of *hearing in*. In the same way, Zak does not just hear a cover version of 'Be My Baby' and think of the original. Rather, he hears the original in the cover.

Leddington coins the term *pictorial cover* for a cover version where the artist intends for the audience to hear the original in the new version. As he writes, 'in many cases, we can— and are *meant to*— hear the canonical track in the cover, and this constitutes a good bit of our aesthetic interest in it, even if the cover is also independently musically interesting.' He offers the example of Stevie Ray Vaughn's instrumental cover of Jimi Hendrix's 'Little Wing' and suggests that 'anyone familiar with the Hendrix will hear both its guitar and vocals in the Vaughan; and much of the pleasure we take in listening to the latter lies in appreciating how it allows us to hear the former in what is a very different piece of music' (2021: 358). A pictorial cover goes further than just referring or alluding to the original. The phenomenal experience of listening to the cover is meant to be different. The audience is not just meant to recognize connections to the original, but instead they are asked to hear the cover version differently than they would if they did not know about the original.

Leddington's example of 'Little Wing' is tempting perhaps because Vaughn's version is instrumental, and it is easy to hear in the song's lyrics. This phenomenon can occur with any instrumental cover, but it is not necessarily an instance of hearing in the original. Just hearing the words is hearing the *song*. To hear the original *recording*, one must hear the lyrics as they are sung in the original. In this case, one must hear in Jimi Hendrix's voice. Leddington's claim that *anyone* familiar with the Hendrix original will hear it in Vaughn's version is simply false. I do not. At most, *some* listeners familiar with Hendrix's original hear it in Vaughn's cover.

Taking Leddington's side, one might say that this only shows that I am missing something important about Vaughn's version— that if I were properly attentive, I would hear Hendrix's original in it. Yet that would presume the very thing in contention, that Vaughn's version is a pictorial cover. Leddington does not cite any declarations from Vaughn which show he meant for his audience to hear Hendrix's version in his. Leddington hears Hendrix in Vaughn's version, others of us do not, and it is hard to carry the point any further than that.

Even though I am unconvinced by that example, Leddington is right that there are pictorial covers. Consider a group of musicians hired to play as a cover band in a bar. The management is paying for there to be live music, but the crowd wants to hear familiar songs. The band is not performing mimic covers, but they are performing fairly straight renditions. The songs are ones that the audience is likely to be familiar with, and the band wants to evoke the originals for the audience. As one musician who used to play in such bands comments: As a crappy bar band, you want any help you can get. So not only does the audience *hear in* the originals, the band wants them to do so.

Something similar holds for more sophisticated cover bands such as Postmodern Jukebox, Scary Pockets, and Tim Akers & The Smoking Section. These groups post videos of some of their covers to the internet, and people who share the videos on social media often do so precisely because of their fondness for the original version of the song being covered. Part of the charm of the renditions is that one can hear the originals in them.

Reference and appreciation

The semiotic angle on covers suggests two different but ultimately related ways of denying the two-mode account of evaluating rendition covers:

- **Impossibility:** Covers (or at least many covers in a sizable class) can only be *evaluated at all* by considering them in relation to the canonical recordings of which they are covers.

- **Incompleteness:** Covers (or at least many covers in a sizable class) can only be *fully appreciated* by considering them in relation to the canonical recordings of which they are covers.

Kania writes that 'it is impossible to properly appreciate an allusion without considering what it is an allusion *to*' (2020: 239). Regarding covers especially, he concludes: 'A rendition cover may be a saturated allusion to its original or not. If not, it can be fully appreciated independently of the original track; but if it is a saturated allusion, it cannot be fully appreciated without comparison to the original' (2020: 243). Emphasizing the former yields the claim that it is impossible to properly appreciate a saturated allusion cover without considering the original it alludes to— that is, Impossibility. Emphasizing the latter yields the claim that one cannot *fully* appreciate the allusive cover without considering its relation to the original— that is, Incompleteness.

Leddington concludes that 'fully appreciating a pictorial cover requires the sort of familiarity with the canonical track that allows you to hear it in the cover. Just *knowing* that it is a cover is not enough. If you cannot hear the canonical track in the cover, then all you hear are the cover's surface features, and you are auditorily and aesthetically missing out on something essential about the work' (2021: 358). Emphasizing the first sentence (*'fully* appreciating a cover'), this sounds like a claim of Incompleteness. Emphasizing the last sentence ('missing out on something *essential*'), this sounds like a claim of Impossibility.

So both saturated allusion covers and pictorial covers can be taken to ground either thesis. In the next several sections, I defend the two-mode account of appreciating rendition covers against both challenges.

Note that both Impossibility and Incompleteness are about all covers or at least all those in a large category, so they cannot be established by offering just one or a few examples. I think they are both wrong as *general* claims. However, there are some specific rendition covers which ought not be evaluated on their own— ones which really should be considered in relation to the recording which they cover. This will depend on the details of the specific case, though, rather than following simply from being a cover or from a specific intention that the covering artist has. (I return to this point below.)

Against Impossibility

There are two general considerations which defeat Impossibility. First, even the most referential of covers can still be appreciated immediately— for example, for its beauty— without considering the original. (This applies one of the lessons of Chapter 3.) Second, Impossibility relies too much on musicians' intentions.

I pursue these responses in the next two sections. Keep in mind that this takes us several turns into the dialogue: I offered the two-mode account of appreciating renditions. Impossibility is a challenge to that account. I will advance responses to Impossibility (thereby defending the two-mode view). In places, I will consider possible objections to my responses, and I will answer those objections.

The value of immediate experience

It really seems as if, arguments notwithstanding, Impossibility is a non-starter. There is a straightforward sense in which one can make an authentic aesthetic judgement about any rendition cover without considering the canonical recording. You can listen to it and find it to be beautiful, ugly, inspiring, saddening, harmonically rich, singable, danceable… and so on. These are reactions to the cover version itself. They are the kind of reactions one might have if one were ignorant of the fact it was a cover or if one were unfamiliar with the canonical version. It is often possible to have these kinds of reactions even after you learn it is a cover and become familiar with the canonical version.

So the claim of Impossibility requires that these categorical aesthetic judgments are somehow illegitimate or confused. Even though understanding an allusion *qua* allusion requires knowing about the earlier recording that the cover alludes to, it is unclear why finding a saturated allusion cover to be beautiful would necessarily implicate the earlier recording.

For a pictorial cover, however, one might argue that finding the cover beautiful without considering the canonical version encounters the cover in the wrong way. The argument might go like this: Imagine someone views a large pointillist painting but stands so close that they just see distinct dots and are unable to resolve it into a picture. If they judge the painting to be beautiful under these circumstances, they have not really judged it properly. They have evaluated the wrong object. Similarly, if someone listens to a pictorial cover without hearing the original in it, then the beauty they attribute to it is misplaced.

This objection requires that failing to hear the original in a pictorial cover is a tremendous mistake, on the scale of only seeing the dots of a pointillist painting. Other mistakes do not nullify aesthetic judgements in the same way. Some examples: One might mistakenly hear the clavinet in Stevie Wonder's 'Superstition' as guitar. One might mistakenly hear 'qu'est-ce que c'est?' and other French phrases in Talking Heads' 'Psycho Killer' as nonsense syllables. Those mistakes would change the phenomenal experience somewhat, because guitars and nonsense will activate different expectations and conjure different meanings than clavinets and French. Nevertheless, I think one can make legitimate judgements of beauty when making those mistakes. Recognizing the instrument or the language would allow one a *better* appreciation of those tracks, but it is not required for there to be any legitimate appreciation at all. If the mistake of not hearing the original in a pictorial cover is like those cases, then pictorial covers support (at most) Incompleteness rather than Impossibility.

The mystery of intentions

A separate problem for Impossibility is the role that intentions play in the accounts of saturated allusions and pictorial covers. Following Stephanie Ross (1981) and William Irwin (2001), Gracyk thinks that allusion requires intention. He writes that, for a saturated allusion cover, the artist 'must intend to communicate with a particular audience… and must intend to have the remake interpreted as referencing and replying to the earlier interpretation' (2012/3: 25). Kania writes similarly that 'not all renditions are saturated allusions— that depends on the intentions of a given track's creator' (2020: 240). Pictorial covers are also defined in terms of the artist's intentions.

Gracyk gives the example of Bob Dylan performing a song by Charles Aznevour and transcribes the monologue that Dylan gives introducing the song (2012/3: 27). As he says in another context, the 'intentions of others are known by reference to their behaviors and statements' (1996: 29). Dylan says enough that we know what his intentions are. However, artists are not always so explicit. Consider Sid Vicious singing 'My Way', which I argued is an implicit reference to Frank Sinatra's version of the song. Appel suggests leaving the matter to 'punk historians', but there may be no evidence that would settle the matter (2018: 449). Vicious, unlike Dylan, is unlikely to have provided an eloquent description of his intentions.

Even where there are strong suspicions, they may be wrong. Ray Padgett discusses the work of Juliana Hatfield, who has a 'rich and varied side gig in tribute albums.' Why had she recorded so many covers? Padgett writes, 'I thought I knew the answer: She adored the artists she was paying tribute to. As I promptly learned, I thought wrong' (2020: 77). In his interview with Hatfield, she reveals that her motivations varied from the desire to record with a friend of hers, the chance to meet a legendary producer, to mere whim. Given her intentions, she would have participated in the projects even if she did not know they were covers or if there had not been an earlier version.

A further wrinkle is that artists might not even be aware of their own intentions. Ross allows for the possibility that the intention underlying an allusion might be unconscious. This means that there might be an allusion even if the artist sincerely denies having intended it. They might be unaware of their unconscious intention (Ross 1981: 61). Irwin allows for unconscious intentions, too, especially in cases where an artist is unaware of an allusion but— when it is pointed out to them— claims to have intended it (2001: 291). So one could say that, regardless of her suggestions to the contrary, Hatfield unconsciously intended to refer to the earlier versions with her covers. If there are no limits to this maneuver— if one can attribute unconscious intentions anywhere— then appeals to intention become frivolous. Discussions of allusion suggest two plausible constraints.

As a first constraint: Attribute an intention when the interpretation of the work itself supports it. If a character in a short film smokes a pipe, wears a deerstalker cap, and draws clever inferences, then it is an allusion to Sherlock Holmes. We would think that even if the creator of the film denies that it is a reference to Holmes. Unfortunately, this is no help for Impossibility. If we attributed the intention to refer only on the basis of interpretive and appreciative considerations, then of course such attributions would obey appreciative constraints. The Impossibility thesis would be established as true, but at the cost of making it a tautology.

As a second constraint: Do not attribute an intention if the artist did not know about the object of the would-be intention. Ross maintains that 'unconscious intent is plausible only when additional evidence… shows that the artist had some knowledge or experience of the work alluded to' (1981: 61). Gracyk makes a similar move (2007: 71). This might help in the case of the filmmaker. Although it is hard to imagine, if she had simply put together the pipe, the deerstalker cap, and crime-solving behavior haphazardly— with no knowledge of Sherlock Holmes— then she must not have been alluding to him after all. Although this might be helpful with other allusions, it is no help in thinking about covers. Artists who make covers necessarily know about the earlier recording, in order to make any kind of cover. So the mere fact of their knowledge cannot settle whether the cover is a saturated allusion or not.

One might attempt to resolve this uncertainty by saying that social conventions and background assumptions can stand in the place of private intentions. As Gracyk writes, our intuitions should be informed by 'the practices of the tradition' (1996: 24). If we judge intentions in this way, recordings for Embassy records and *Top of the Pops* were mimic covers because of the organization of those publishers regardless of what was going on inside the heads of individual session musicians. Similarly, one might say that recording songs for tribute albums makes Hatfield's covers saturated allusions, regardless of what private intentions she had. However, the practice of recording tribute albums does not support such an assumption. Although some tribute albums are intended for listeners who are familiar with the original versions and so could recognize the reference, others are intended for a new audience that is unfamiliar with the classic songs. For example, the 1991 Leonard Cohen tribute album *I'm Your Fan* was recorded with the hope of making a younger audience more familiar with Cohen's work (Padgett 2020).

One might concede all of the difficulties figuring out intentions, but insist that intentions nevertheless exist. The idea would be that, even if they are unknown, intentions might still determine what kind a cover is and what kind of appreciation is appropriate. Gracyk can be read as offering such a metaphysics-first approach (1996: 26). However, there are not always well-defined or determinate intentions. An exhausted or drug-addled musician might not have any defined intention about how their work should be understood. Alternately, an artist might have an open-ended intention. They might want listeners both to listen to their version as its own thing and to consider it in relation to the original, or they may be happy for audiences to encounter it in either way. Attributing complex intentions to artists may be appropriate in some cases, but in other cases it over-intellectualizes the whole process.

There is also the extra complication of multiple people involved. The singer, other members of the band, the producer, and the A&R might all have different intentions for how audiences should react to a cover. As Gracyk himself notes in another context, 'rock recordings frustrate the expectation that each work features *an* artist's intentions' (1996: 94). Because there is not one governing intention, there will often be no definite fact of the matter about how a cover is intended to be heard. Yet being a saturated allusion or pictorial cover depends on how the cover is intended to be heard, and the arguments for Impossibility started from those kinds of covers. The ambiguity of intention makes those categories too unstable to support such a weighty conclusion.

So the two-mode account of evaluating renditions survives the challenge of Impossibility.

Against Incompleteness

Recall that the claim of Incompleteness is the idea that, although a saturated allusion or pictorial cover can be appreciated separately from the canonical recording of which it is a cover, it can only be *fully appreciated* in relation to the canonical recording. Whether this is true or not will depend importantly on what 'fully appreciated' means.

One might understand full appreciation to be *complete*— that is, appreciation in all the ways. This would make Incompleteness both trivially true and irrelevant. It would be trivially true because, if you have not yet appreciated the cover in relation to the original, then you have not appreciated it in all the ways yet. And it would be irrelevant because it would be no objection to the two-mode view of evaluating rendition covers. My claim was not that there are two ways to *fully* appreciate a cover, after all, but just that there are two ways of coming at it. This is compatible with (even entails) that consideration in just one mode is partial. So understanding Incompleteness as something besides a trivial *non sequitur* will require understanding full appreciation in some other sense.

Another way of understanding it is that fuller appreciation would consider the *deeper, more significant* things. The idea would be this: With a saturated allusion cover or pictorial cover, one is meant to consider the cover in relation to the original. So that aspect is more central and significant than its superficial features in a way that is effectively essential to full appreciation.

One problem with this is that knowledge of the original might not actually be more central and significant. Historical factors and references may be important variables for appreciating a work, but sometimes not. There are cases where hearing the original does not add anything of artistic or aesthetic value to appreciation of the cover, beyond the knowledge that the song is not an original composition. One might still think that a trivial addition to one's appreciation would be *some* addition, but that falls back on the sense of 'fuller' as more complete.

Of course, judging the originality of the cover requires considering how it differs from the canonical recording. However, as I discussed in the previous chapter, assessing originality may also require knowing about other recordings, other songs, other musicians— anything that might be a source of inspiration for the cover version. As Gracyk notes in a different context, 'Philosophically, once you've made the move toward a historically-aware, contextualist understanding of both aesthetic judgment and artistic value, it's a short step to the idea that any and all aspects of the social context of production and reception might be relevant to the music's characteristics and value. (I stress "might be" here: any and all might be, but in any given case only some will be.)' (2021). If one insists that the cover's connection to the original is somehow the essential lynchpin to appreciating it, then one has reverted to the claim of Impossibility. If it is not strictly essential, then facts about the original become just some of the many historical facts which might— or might not— be relevant to appreciating the cover version.

This problem is compounded by the fact that considering the cover in relation to these additional facts and information may undercut appreciating it in other ways. It may simply not be possible (for some listeners at least) to hold both modes of evaluation in their head at once. So considering a cover in one mode would preclude considering it in the other. This might occur when the quantity of information would change the focus and mood. Reflecting on the beautiful simplicity of a particular instrumental might conflict with having lots of contextual facts in mind. When someone complains that thinking too much about a song ruins their experience of it, it would be elitist snobbery to simply deny that they were having the best experience of it anyway. (That sentence could be the pull quote for the book!)

Moreover, once someone experiences the original, they may be unable to hear the cover in the same way. If they *hear* an original's lyrics *into* an instrumental cover, they may lose the ability to hear it as just an instrumental. If they hear the earlier version into a cover that changes the genre, the contrast might overwhelm their ability to hear the song in its new setting. Hearing the pop original of a punk cover, for example, might change what features stand out to them. Jesse Prinz, paraphrasing Matthew Kieran (2008), suggests that 'becoming a punk enthusiast can diminish one's tolerance for other genres, making them seem overproduced, tame, or vapid' (Prinz 2014: 589). And listening to the pop original of a punk cover might highlight tame, vapid features that survive as traces in the punk version, leaving the listener with a worse experience than they had before. It is possible that these vapid elements, once heard, cannot be unheard.

This is especially an issue when the experience of the cover involves hearing in the original. When Albin Zak hears a cover of 'Be My Baby', irresistably 'the original recording still comes through' (2001: 31–32). There might be covers which are so jarringly different that Zak cannot appreciate them, where the contrast with the original will always overwhelm whatever charms the new version might offer to a fresh ear.

So Incompleteness either collapses into triviality (by requiring all information for complete appreciation) or is tantamount to Impossibility (by making consideration of the original essential to appreciation).

Further reflections on allusions

Although I think the arguments above suffice to defeat Impossibility and Incompleteness, I want to return to a related issue about allusions. A crucial premise about saturated allusion covers was Kania's claim, quoted several times above, that 'it is impossible to properly appreciate an allusion without considering what it is an allusion *to*' (2020: 239). This claim is not true of all allusions, which shows that allusions will not carry the weight that Gracyk, Rings, and Kania are placing on them. I think this may be partly obscured by the oddity of *saturated* allusions, so let's take a step back and consider how the usual, unsaturated kind of allusion works.

Kania offers the example of the allusion to Vladimir Nabokov's *Lolita* in the Police's 'Don't Stand So Close to Me' (2020: 238). The song tells the story of a 'young teacher' who sleeps with one of his pupils (or who is at least tempted to do so). In the third verse, 'It's no use, he sees her / He starts to shake and cough / Just like the old man in / That book by Nabokov.' A lot is conveyed in the allusion here. *Lolita* is a complicated and controversial novel about a middle-aged professor's molestation of a teenaged girl. The character in the novel is a monster who meets a bad end. The line is the strongest suggestion that the teacher in the song is himself a monster. Not only would I miss all that if I did not know about the book in question, but the line 'That book by Nabokov' would have no significance to me at all.

It is more typical for allusions, especially in popular art, to serve a dual purpose. As Gracyk writes, 'a good allusion will serve as a functional element… independent of its alluding function' (2007: 76). In the song 'It's My

Life', Jon Bon Jovi sings, 'My heart is like an open highway / Like Frankie said, "I did it my way".' This is an allusion to the song 'My Way', written for and canonically sung by Frank Sinatra. The allusion can be seen as an acknowledgement by Bon Jovi, to listeners who might already be thinking it, that the sentiment of 'It's My Life' is much like the sentiment of 'My Way.' For listeners who do not follow the allusion, Bon Jovi is still actually saying that he did it his way— while acknowledging that it is not entirely original to do so, because some other guy already said that. I guess I appreciate Bon Jovi's song better for understanding the allusion, but the non-allusive function of the line is at least as important.

Other cases go further, so that the allusive function is clearly less important than the plain function of a line. Consider two examples from songs by They Might Be Giants (TMBG). First, the line 'I Hope That I Get Old Before I Die' (from their song of the same name), an allusion to the line 'I hope I die before I get old' from the Who's 'My Generation.' Second, the lines 'Quit my job down at the car wash / Didn't have to write no one a goodbye note' (from their song 'Put Your Hand Inside the Puppet Head'), an allusion to 'Guitar Man', a song written by Jerry Reed and made famous by Elvis Presley, which has the lines 'I quit my job down at the car wash / Left my mamma a goodbye note.' In both of these cases, the lines in TMBG's songs work even for listeners who do not recognize them as allusions. There may be some small pleasure or appreciative value in recognizing the allusions, but it clearly does not eclipse the lines' non-allusive face value.

Other allusions are musical rather than lyrical. Recall the eighth-note handclaps in the Beatles' cover of 'Words of Love.' Zak argues that the allusion there is important, because it connects the cover not just to one Buddy Holly song but to Buddy Holly's sound in general. However, musical allusion is not always so weighty. Consider the cover of Weird Al Yankovic's 'Eat It' by the Japanese punk band Shonen Knife. The bassist opens and closes the song by playing the bass line from Deep Purple's 'Smoke on the Water.' It fits and sounds good in the Shonen Knife track, and it offers a little something extra for listeners who recognize it. It works like quotation often does in jazz performance, where a soloist quotes a short passage from another jazz standard or a famous improvisation. This can be an homage or an in-joke, but it is intended to serve as a coherent part of the solo. In musical quotation, as in the lines from TMBG songs, there is no explicit statement that tells the listener that the passage is an allusion.

So Kania's example of the line from 'Don't Stand So Close to Me' is peculiar in two respects: First, the line only serves an alluding function rather than having a dual purpose. The more usual kind of allusion can be appreciated either for its function in the song or for its alluding function. Neither is blocked in principle, even if one mode of appreciation might be unrewarding in a particular case. Second, the line about Nabokov is explicit about being an allusion. Many allusions are quotations without explicit markers that quotation is happening.

For a saturated allusion cover, the entire cover version is intended to refer to the canonical original. That makes it less like the line about Nabokov and more like the other examples. First, since the entirety of the song is performed or recorded for the cover version, a listener can attend to all of its features without thinking of the original. The cover serves the face-value function of being a version of a song in addition to the alluding function of referring to the original. Second, except in the case of explicitly referential covers, the connection to the earlier version is not explicit. It is more like quotation than it is like the Police's reference to 'that book.'

By analogy with particular allusions, then, a saturated allusion cover can be appreciated either on its own or in relation to the recording of which it is a cover. That is to say, the two-mode account of appreciating rendition covers applies.

Evaluating rendition covers, revisited

So far I have defended the two-mode account of appreciating rendition covers, but I think that the discussion above allows us to extend the account a bit. There are two ways one can go about appreciating a cover in relation to the earlier recording it covers: One can have facts about the canonical recording in mind, or one can hear the canonical recording in the cover version. The latter involves greater changes the experience of the cover itself, shaping the implicit anticipations which alter the overall perceptual gestalt. Because *hearing in* amounts to hearing differently, it can make it hard to hear the cover as it sounds to an uninformed listener. So appreciating a cover by hearing in the original may preclude appreciating it in other ways.

This yields what one might call two and a half modes of appreciation. Although in principle all are possible in every case, they will not all be rewarding or worthwhile in every case. Moreover, fully engaging with some might preclude engaging with others.

In this section and the next, I want to consider how these factors play out for some particular covers.

Consider Johnny Cash's 2002 cover of 'Hurt.' The song was written by Trent Reznor and originally released in 1994. After seeing the music video for Cash's version, Bono commented, 'Trent Reznor was born to write the song. but Johnny Cash was born to sing it.' Reznor himself said, describing his reaction to the video, 'Tears welling, silence, goose bumps... that song isn't mine anymore' (Padgett 2017: 207). For those of us familiar with the original, Cash's cover was surprising. However, the Johnny Cash version is now the canonical version of the song for some listeners. Although Cash was a songwriter, he also regularly recorded songs he had not written. One can appreciate most of those without reference to any earlier recording. For example, one can listen to Cash's version of 'Ghost Riders in the Sky' and treat it as part of the country music repertoire, without knowing anything about Stan Jones (who wrote it) or having heard Burl Ives' 1949 version (the first released recording). One can listen to Cash's 'Hurt' in the same way. However, Reznor's song is neither country music nor part of a standard repertoire. It is deeply personal and not— or at least not before Cash— an obvious thing to cover. So it is also rewarding to consider Cash's version in relation to Reznor's original.

As a contrasting example, consider Eric Clapton's 1974 cover of Bob Marley and the Wailers' 'I Shot the Sheriff.' Gracyk counts it as a mere remake rather than a saturated allusion on the grounds that 'Clapton was not comfortable with reggae and did not want to record the song or release it, but was urged to do so by his band mates and producer' (2012/3: 27). Clapton is neither trying to sound like Marley (so it is not a mimic cover) nor referring to Marley (so it is not an allusion or a picture). *Billboard* included Clapton's version among its Top Single Picks with no comment on it being a cover, calling the track 'a catchy goof of a winner.' The staff writers describe it as having 'the latino percussiveness and broad outlaw storyline of "Cisco Kid"' and add that one 'reviewer found himself humming it 11 hours straight' (1974). This appreciation of the Clapton track without consideration of the original is deeply impoverished. What the Billboard reviewers hear as 'latino percussiveness' can be heard instead as the residual reggae influence from the original. Their capsule review reveals that, although the cover can be considered in isolation from the original, it is misleading to consider it that way. The cover is more profitably appreciated in relation to the original version, and I suspect that someone who has heard the original will no longer be able to hear the 'latino percussiveness' that appeared to naïve listeners.

Note that this example does not revive the claim of Impossibility. Because Clapton's cover makes no implicit reference to the original, it is not part of the class of covers that Impossibility was supposed to apply to. In fact, Gracyk holds both that a saturated allusion cover should be considered in relation to the original and also that a non-allusive cover (what he calls a mere remake) should be considered just as an instance of the song. Clapton's version is non-allusive but nevertheless is best considered in relation to the original. So it provides more evidence for the two-mode view— that any rendition cover may *in principle* be approached in either way. In cases where only one approach is rewarding, it is because of the particular artistic and aesthetic details rather than because of any general rule.

Covers of covers

The modes of evaluation proliferate even more when a version is a cover of a cover. In the schematic case, imagine an original recording— call it #1. Another artist records a rendition cover of #1— call it #2. A third artist records another rendition, clearly respecting some of the arrangement and musical choices of #2— call this #3. In evaluating #3, one might consider it on its own, in relation to #1, in relation to #2, or in relation to both. And the latter modes might be with or without hearing in. In the abstract case, this yields at least seven possibilities. Just as for the simpler case of a cover of one original, however, not all the theoretically possible modes will be rewarding or worthwhile— and it may not be possible to pursue them all.

Let's consider several examples.

First, consider college a cappella groups covering 'Bitches Ain't Shit.' The song was written and released by Dr. Dre in 1992, but a cappella groups more clearly follow Ben Folds' 2005 cover. If we consider it in relation to earlier versions, it is best to do so in relation to *both* the original and Folds' cover. The lyrics are due to Dre, but the surprising genre shift is due to Folds. I can easily hear Folds' version in the a cappella performances, but I am hard-pressed to hear Dre's original in them. (For more details about this example, refer to the discussion of it in Chapter 3.)

Second, consider Jeff Buckley's well known cover of Leonard Cohen's

'Hallelujah.' Initially at least, Buckley only knew the song from John Cale's cover of it. Although 'Hallelujah' appeared on Cohen's 1984 album *Various Positions*, his record label was so skeptical of the album's prospects that they did not release it in the United States. Cale, having heard the song at a live performance in New York, decided to record it for the 1991 tribute album *I'm Your Fan*. Cohen sent Cale the written lyrics, which included many more verses than Cohen's recorded version. Cale did not use all the verses he was sent. Cale settled on five verses, only two of which are shared with Cohen's earlier recording. Buckley's version replaces the piano part from Cale's version with guitar, but mostly follows Cale's lyrics— unsurprisingly, since Buckley had neither heard Cohen's version nor seen Cohen's written lyrics. The expressiveness of Buckley's version is largely due to his vocal stylings. The song was commonly performed as a tribute to Buckley after his untimely death. Although the song has been covered by numerous artists, Buckley's version has often been considered definitive. One might listen to other versions if one wants to evaluate the song— and one might find the historical path of the song to be an intriguing tale— but I think that appreciating Buckley's version is neither richer nor more rewarding for hearing earlier versions. (For more details, see Light 2012 and Padgett 2020.)

Third, consider 'Hound Dog': a song canonically associated with Elvis Presley. The original track, recorded by Big Mama Thornton, is about a no-good, cheating man. Where Elvis' version has the lines 'Well, you ain't never caught a rabbit / And you ain't no friend of mine', Thornton's has 'You can wag your tail / But I ain't gonna' feed you no more.' This revision was not original with Elvis. It was first made by Frankie Bell and the Bell Boys, who used their silly version of the song as the closing number of their Las Vegas act. That is where Elvis encountered it (Padgett 2017: 15). It is possible that Elvis never even heard Thornton's version. Nevertheless, I do not think it adds much appreciatively to learn about Bell. Hearing Bell's version into Elvis' version would not create a richer or more rewarding experience. Considering Elvis' version in relation to Thornton's does change matters, though. One might think, as cowriter of the original Jerry Leiber did, that Elvis' version 'ruined the song' by turning 'a song that had to do with obliterated romance' into 'inane' nonsense (Padgett 2017: 23).

Fourth, consider 'Killing Me Softly.' The lyrics are based on a poem by Lori Lieberman, who recorded and released the song as 'Killing Me Softly With His Song' in 1972. Her version was heard by Roberta Flack, who rear-

Some Covers of Covers	Is it rewarding to consider it in relation to the original?	Is it rewarding to consider it in relation to the earlier cover?
a cappella "Bitches Ain't Shit"	✅	✅
Jeff Buckley's "Hallelujah"	❌	❌
Elvis' "Hound Dog"	✅	❌
The Fugee's "Killing Me Softly"	❌	✅

Figure 8: This table shows different cases discussed in the chapter. Each highlights a different appreciative possibility. If you disagree with where I have placed some of these, consider other covers of covers to find your own examples.

ranged the song and released a version in 1973. Flack's version became well known and was the canonical version in 1996 when the Fugees released a cover of it. It is plausible to think of the Fugee's cover as picturing Flack's track, because Lauryn Hill of the Fugees recorded thirty separate harmony parts to reflect the background vocals in Flack's version. Few listeners to either Flack's or the Fugees' versions even know about Lieberman's version. The contrast with Lieberman's original may contribute to appreciation of Flack's cover, underscoring the originality of Flack's reinvention of the song. However, Lieberman's original is irrelevant to appreciating the Fugees' cover of Flack.

Coda

This chapter began with a focus on covers which refer in some way to earlier canonical recordings. Whether by explicit mention, saturated allusion, or picturing, this reference allows the cover to comment on the canonical version. Such a cover might raise issues with the earlier recording which would not have occurred to someone who listened to the original but had not heard the cover. It is even possible to hear a cover into an original, so that the experience of hearing the original is different once you know about the cover.

For example, I was familiar with Jimi Hendrix's cover of 'All Along The Watchtower' before I heard Bob Dylan's original. When listening to Dylan's original, I tend to hear Hendrix's version in it. Of course, the original is not pictorial— Dylan could not have foreseen Hendrix's cover, so he could not have intended for listeners to hear it in his recording. Nevertheless, I notice features of the original that I would not be able to appreciate if it were not for that connection.

Consider also 'Iron Man', originally by Black Sabbath and covered by the Cardigans. From the Cardigan's cover, I have learned to hear a certain sadness in the song. I can hear that sadness when listening to Black Sabbath's original version, but only because I heard it in the Cardigan's first. Without having heard the cover, I would only have noticed the anger and madness.

Because a cover can serve as commentary on the original, it can reveal features of it that one would not notice otherwise. To put this in a deliberately provocative way: Sometimes, one cannot fully appreciate an original without hearing the covers of it.

Interlude: Torment and Interpolations

There may come a time— if it has not been reached already— when all the great works of music have been written. John Stuart Mill writes: 'The octave consists only of five tones and two semi-tones, which can be put together in only a limited number of ways, of which but a small proportion are beautiful: most of these, it seemed to me, must have been already discovered, and there could not be room for a long succession of Mozarts and Webers, to strike out, as these had done, entirely new and surpassingly rich veins of musical beauty' (1873: ch. 5).

Mill's idea is that there are only a finite number of notes which can be combined in only a finite number of ways. Many of the ways will be awful. Of those that are not, many have already been discovered and documented by great composers.

Mill recounts that he was 'seriously tormented' by this line of thought, but notably this confession is in his autobiography rather than in one of his philosophical works. He makes light of it in retrospect, offering it as evidence of the dark place he was in rather than suggesting it as a real concern. The torment, he writes, was 'very characteristic both of my then state, and of the general tone of my mind at this period of my life.'

In order to make it a cause for concern, we would need the additional assumption that running out of tunes would be bad. Mill suggests that 'the pleasure of music… fades with familiarity, and requires either to be revived by intermittence, or fed by continual novelty.' For many of us, however, there are favorite songs which can survive being replayed. Even overly-familiar songs can be given new life by a new musician who changes them up. That is part of the fun of covers.

The argument also relies on a questionable assumption of 'the exhaustibility of musical combinations.' That is, it requires that the palette of musi-

 https://doi.org/10.11647/OBP.0293.11

cal materials is sufficiently limited that musicians might explore the whole space of worthwhile combinations. Even if the number of possible songs is finite, it might still be so large that musicians would not write all the good songs even in the lifetime of the whole universe. Moreover, musical performance can offer an uncountable infinity of qualities— in timbre, timing, and expression— such that the same musical passage can offer different rewards when played by different musicians.

Nevertheless, Mill is right that there are only so many ways to put together a finite set of notes and chord progressions. Given the structure of a pop song, there are only so many possible melodies, choruses, or bridges. So it is no surprise that many patterns appear in multiple songs. Earlier songs often serve as inspirations for new ones, and songwriters reuse elements from earlier work. Moreover, it is not unheard of for a songwriter to independently hit upon a melody that has already been used by someone else.

The industry term for using the melody from a copyrighted song is *interpolation*. Here is a typical definition: 'Interpolation is when you use any portion of lyrics or melody from a copyrighted song that you did not write...' (Easy Song 2021). It is easy to think of interpolations as being almost but not quite covers. For example, Adam Neeley comments on a particular interpolation, 'In the eyes of the law it's not a cover, but it's also not a wholly original song' (2021).

However, it is important to note that the law does not specify what it means to be a *cover*. Typically, legal decisions do not even turn on whether two versions are the *same song*. Rather, what matters is only whether they are similar enough that the later one steps on the copyright of the earlier one.

Moreover, an interpolation— unlike a cover or quotation— need not be deliberate. The upshot of Mill's argument, combined with the relative simplicity of pop music melodies, is that interpolations will happen by accident. For example, Sam Smith's 2014 hit single 'Stay With Me' has a melody and chorus with 'notable similarities' to Tom Petty and Jeff Lynne's 1989 'I Won't Back Down.' News coverage indicates that 'it wasn't a deliberate thing' but instead a 'complete coincidence.' Nevertheless, when matters were settled, Petty and Lynne were added to Smith's song as cowriters. Petty issued a statement saying, 'All my years of songwriting have shown me these things can happen. Most times you catch it before it gets out the studio door but in this case it got by. ... A musical accident no more no less' (Coplan 2021).

As Hannah Sparks comments, 'It's not uncommon for today's superstars to retroactively credit additional writers, thus dealing them in for potential royalties' (2021). I could add further examples, but— for reasons Mill anticipated— it is inevitable.

5. Some Metaphysical Puzzles About Songs

The Dead Kennedys' 1979 track 'California Über Alles' begins with Jello Biafra singing the line 'I am Governor Jerry Brown', and a line about 'Carter power' refers to President Jimmy Carter— that is, the lyrics namedrop then-current leaders. A verse about secret police and death camps begins 'Now it's 1984', alluding to George Orwell's *1984* but also looking a few years ahead. The track was successful and has been covered numerous times. A few examples include recordings by Six Feet Under in 2000, The Delgados in 2006, and Vio-Lence in 2020. These later bands sing the same words that Biafra sang in 1979, but the lyrics are out of place. They are a relic of a California long passed, but the covering bands sing them because they want to record the same song that the Dead Kennedys recorded.

The Disposable Heroes Of Hiphoprisy's 1992 cover of 'California Über Alles' both changes the genre (from punk to hip-hop) and changes the lyrics. It begins with a sample of Biafra singing the title, and then Michael Franti of the Disposable Heroes sings 'I'm your governor Pete Wilson.' There's no mention of 'Carter power', and the verse about secret police begins with 'Now it's 1992.' By changing the lyrics in this way, the Disposable Heroes bring the song up to date— but do these changes mean that they are singing a different song?

One might argue: The Disposable Heroes' version of 'California Über Alles' is a cover of the Dead Kennedys' track, so it must be an instance of the same song. This supposes that cover versions are always the same song— a problematic assumption, it turns out. It also fails to capture the important sense that verbatim covers by Six Feet Under and others are importantly different than the Disposable Heroes' version.

One might argue instead: The Disposable Heroes' version is about different people and times than the Dead Kennedys' version, so it is a different

 https://doi.org/10.11647/OBP.0293.05

song. It is clearly based on the Dead Kennedys' original, but that does not make it the same song any more than sampling three words of Biafra's original vocal does.

Or one might argue: 'California Über Alles' is meant to be a politically-charged attack on the governor. Six Feet Under's version is about someone who had not been governor for 17 years, so their version lacks the right political valence to be an authentic instance of the song. It is 'California Über Alles' in name only. (The original lyrics were perhaps relevant again in 2011–2019, during Jerry Brown's second stint as governor.) Pursuing that line of reasoning, the Disposable Heroes' version is the real thing because it called out the then-current governor.

Let's put this specific example aside for a moment. I will not be able to offer a solution until the next chapter. The rest of this chapter raises other, related concerns about what it takes for two versions to count as instances of the same song.

Interpolated covers

In pop music, the word 'interpolation' is often used to contrast with *cover*. It means a version which uses parts from an earlier song, rather than one which uses so much as to be the same song.

In rap and hip-hop, 'interpolation' is used to contrast with *sampling*. It means rerecording a vocal or instrumental part of an earlier track instead of using a sample. For example, Wu-Tang Clan was able to secure rights to the Beatles' song 'While My Guitar Gently Weeps' but not to the recording of George Harrison's original guitar part. So they interpolated it by having Dhani Harrison play the guitar part in the studio (Montgomery 2007). In some hip-hop tracks, the interpolation is just an isolated fragment. In others, like Wu-Tang's 'The Heart Gently Weeps', the chorus or refrain from the earlier recording is used as the hook which connects rapped verses with new lyrics. Claire McLeish coins the term *interpolated cover* to describe hip-hop tracks of this form (2020: ch. 5). Examples which McLeish discusses include three from 1988: The Real Roxanne's 'Respect' (covering Aretha Franklin's 1967 hit), the Fat Boys' 'The Twist (Yo, Twist!)' (covering Chubby Checker's 1960 hit), and 2 Live Crew's 'Do Wah Diddy' (covering Manfred Mann's 1964 hit). These have the same or similar titles to the earlier recordings as 'an easy way for hip-hop artists to indicate which earlier songs inspired their own' (2020: 195). McLeish takes interpolated covers to be almost but not quite the same song— more similar than a plainly new song but less similar than an ordinary cover.

Another example of an interpolated cover is Hilary Duff's 2008 track 'Reach Out', which changes lyrics and adds rapped sections to Depeche Mode's 1989 'Personal Jesus.' In Depeche Mode's original, the invocation of Jesus and the demand to 'Reach out and touch faith' are a metaphor for obsessive love. Riley Haas describes it as having 'sex appeal with a sinister undercurrent of dominance and submission' (2020). In Duff's version, where the chorus is 'Reach out and touch me', the themes of sex and submission are all on the surface. Dan Burkett identifies Duff's version as a cover and writes that fans recognize Depeche Mode and Duff as 'performing the *same rock songs*' (2015). I am not sure whether Burkett is right. Martin Gore (who wrote 'Personal Jesus') is credited as one of the writers on 'Reach Out', but there are two other credited writers. Online discussions of Duff's track typically do not use the word cover, although some do.

It is instructive to contrast Duff's 'Reach Out' with two other tracks. First, consider Johnny Cash's 2002 cover of 'Personal Jesus'. Although Cash sings all the lyrics from the original, the invocation of Jesus in his version is not a metaphor for anything. Cash comments in an interview with Bob Edwards, 'To me it's a very, very fine evangelical song— although I don't think that's why it was written' (Edwards 2002). So, although Cash's version unproblematically counts as a cover, it effaces one half of what is going on in the original just as much as Duff's. Second, consider Jamelia's 2006 track 'Beware of the Dog.' Like Duff's track, it samples the main riff of 'Personal Jesus' and credits Gore as one of the writers. The bulk of 'Beware of the Dog' is straightforwardly a different song, especially when played live— when the riff is played by guitarists rather than being replayed as a sample. Where Duff's version is an interpolated cover, Jamelia's just uses samples or interpolations from 'Personal Jesus' in a new song.

Must a cover be the same song as the original?

In the typical case, a cover is a version of the same song as the original track. For example, They Might Be Giants are singing the same song in their version of 'Istanbul (not Constantinople)' that the Four Lads sang in the original. However, interpolated covers raise the spectre of extraordinary cases:

Are there some covers which are not (versions of) the same song as the recordings that they are covering?

One might argue: No! A cover is a recording of a song that was first recorded by someone else, so the cover and the original are versions of the same song just by definition.

I argued against trying to define 'cover' in Chapter 1, so I think this argument is a non-starter. Even if one decided to stipulate a precise definition of 'cover', that would not help here. Someone else might just as easily stipulate a different definition.

My strategy has been to take the category of so-called *covers* as given, which suggests that maybe what we need is data about how audiences think and talk about covers. Christopher Bartel reports on a number of small experiments that provide some data. Here is a summary of his results:

- 62% of respondents said that a recording of a mimic cover of AC/DC's 'Back in Black' is the same song as the recording of an indistinguishable performance by AC/DC.

- Only 39% said that Whitney Houston's cover of 'I Will Always Love You' is a recording of the same song as Dolly Parton's original. The prompt told respondents that Houston's 'recording contains the same lyrics and the basic melody; but it sounds dramatic, powerful, and heartrending' (Bartel 2018: 358).

- Only 19% said that Johnny Cash's cover of 'Hurt' was a recording of the same song as the Nine Inch Nails original. The prompt told respondents that the instrumentation and some of the lyrics were different in Cash's version, and that 'the song seems to be referencing the aging music legend's failing health' (Bartel 2018: 360).

Bartel was especially interested in the way changes in emotion and meaning affect judgements of whether a cover is the same song, so he constructed the three cases so that the first involves no change, the second involves a change in emotional force, and the third involves a change in meaning. However, it is striking that in *every* case a sizable percentage of participants thought that the cover would count as a different song than the original. Back in Chapter 2, I noted that the word 'song' is used loosely in everyday talk— sometimes it is used to mean the recording. Perhaps that is how Bartel's subjects are using the word. Bartel acknowledges this possibility but notes that, although it would explain how more than a third of respondents counted a mimic cover as a different song than the original, it would not explain the increasing tendency to consider something a different song when there was a greater change in force or meaning (2018: 363). Every cover is a different recording, regardless of whether it is the most faithful mimic or the most transformative rendition.

There are several issues one might raise with Bartel's results: It is just one study. The prompts did not use the word 'cover.' The participants were students in philosophy and music courses rather than experts on music. And so on. Motivated by concerns like these, one might go on to do variants of the study and obtain further results.

I am not going to do that, however. I argued, back in Chapter 2, that my use of the word 'song' is an explication. It is not quite what ordinary people mean by the word, even though it marks a distinction that ordinary people can recognize. More experimental results which show that people do not use the word 'song' this way are just what one would expect. And experimental results which showed how philosophers of music use the word 'song' would just recapitulate the philosophy of music.

Striking covers

The fact that covers can sound different than earlier versions raises what Andrew Kania calls the *striking cover paradox*. The idea is that there could be a series of covers, each making small changes to the one before it, so that the final product sounds nothing at all like the original. As Kania puts it, the outcome could be 'a cover of "Don't Be Cruel" [that] sounds for all the world like "Pop Goes the Weasel"' (2006: 410). Here is the puzzle posed in explicit steps:

The Striking Covers Paradox

Take an original track. Call it A. Someone records a rendition cover of it. Call that first cover B. Someone else records a rendition cover of B. Call it C. And so on for versions D through Z, each a cover of the one before it in the series.

1. Because each is a rendition cover of the one before, each will be at least slightly different than the one before it in the series. Small changes might accumulate so that Z sounds nothing at all like A.

2. Because each track is a cover of the one before it, they are all instances of the same song.

3. It is impossible for two instances of the same song to sound nothing at all alike.

4. Therefore (from 1) it is possible that Z sounds nothing at all like A, and (from 2 and 3) it is impossible that Z sounds nothing at all like A.

The conclusion, that something both *is* and *is not* possible, is an explicit contradiction. The inferential steps seem secure, so there must be a problem with one of the premises.

Kania's resolution to the paradox is to deny step 1. If there were a series such that Z sounded nothing like A, then Kania maintains that at least one of the recordings along the way must not really have been a cover of the one before it. Suppose, for example, that differences accumulated so that Q is the first one in the series that is not an instance of the same song as A. Then Kania would say that Q is not really a cover of P. Note that it might be the case that Q sounds recognizably like P, that the musicians intend for Q to be a cover of P, and that music critics and fans call it a cover. Kania would deny, on principle, that it actually is a cover.

I am open to the possibility that philosophical results can outweigh common usage and practice like this, but it is not a happy outcome. It can be avoided by instead denying step 2. If some covers are different songs than the recordings they cover, then every version in the sequence can be a cover even if the final cover sounds nothing like the first original.

This is all well and good in the abstract, but can we find a real example of a cover version which is not an instance of the same song as the track that it covers?

Regarding crossover versions of doo-wop songs, shorn of their original stylings, David Goldblatt writes that 'for those with the proper sensitivities, the differences were understood to be so great that the two were thought to be the same song only nominally' (2013: 109). Which is to say: The pop covers were not *really* the same song. However, someone taking Kania's position might say that Goldblatt is just speaking figuratively— that is, one might say that doo-wop originals and pop crossovers really are the same song.

Kid Cudi's '50 Ways to Make a Record', a cover of Paul Simon's '50 Ways to Leave Your Lover', refigures the lyrics to make a song about the craft of music. Accepting that it is a different song, someone taking Kania's position might say that Kid Cudi's track is not really a cover.

What we need is an example that resists both replies. There must be a strong case both that it is a different song and also that it is a cover. In work with collaborators, I have used the example of Aretha Franklin's 1967 cover of Otis Redding's 'Respect' (Magnus et al. 2013). It is standardly called a cover, and *Rolling Stone* calls it the 'definitive cover' (2021). However, as Jeff Giles writes, when 'people think of "Respect" — hell, when they just *hear the word* respect — it's Aretha's voice they hear. Through a dizzying blend of flawless technique and raw power, she *owns* "Respect"' (Popdose 2011). Ray Padgett expresses a similar thought when he writes that Franklin 'treated it like a totally new song— which, in many ways, it was' (2017: 50).

Franklin 'transforms Redding's ultimatum to a housebound woman into a demand for consideration, one which might be made between equals' (Magnus et al. 2013: 365). This means that 'Redding and Franklin both sing about respect, but they say importantly different things about it' (366). She does this not just by changing the mood, but by changing many of the melodic, structural, stylistic, and lyrical features. Victoria Malawey details the differences and concludes that 'Franklin re-authors "Respect" to such an extent that ownership transfers from songwriter Redding to Franklin' (2014: 205). I would put the point somewhat differently. It is not that Franklin takes ownership of Redding's song, but instead that— starting from the material of his song— she makes a new one. Among Franklin's changes is the addition of the memorable lines 'R-E-S-P-E-C-T / Find out what it means to me'; Redding never spelled it out.

A further thing to note is that the words in Franklin's version of 'Respect' are not merely changed so that the narrator is female rather than male— rather, the narrator in Franklin's version is understood to be the woman who is being addressed by the narrator in Redding's. Where Redding gives his woman permission to mess around on him when he is away, Franklin says to her man that she has no interest in messing around. So Franklin's cover is not merely in reference to Redding's track, but in dialogue with it.

There is a tradition of answer songs— especially in the 1950s and 1960s— which used the same melody as a popular song but changed the lyrics so as to provide a response. To take just a few examples: Rufus Thomas's 1953 'Bear Cat' was an answer song to Big Mama Thornton's 'Hound Dog' with lyrics from the man's point of view. After Elvis had a hit in 1960 with

'Are You Lonesome Tonight?', Dodie Stevens and Thelma Carpenter both re-leased answer songs titled 'Yes, I'm Lonesome Tonight.' After Dion's 'Run-around Sue' in 1961, Ginger Davis and the Snaps released 'I'm No Run Around.' And so on.

Franklin's 'Respect' has the same title as Redding's, but it can be seen as part of the answer song tradition (Malawey 2014: 196). B. Lee Cooper sur-veys answer songs and notes that, although they are 'usually humorous' and 'regarded as a novelty', 'the functions of specific answer songs vary greatly' (1988: 57, 58). Both of the songs titled 'Yes, I'm Lonesome Tonight' are cheesy love songs in the same vein as the original, rather than being jokes. Franklin's 'Respect' is a serious song about relationships just as much as Redding's.

The song that Franklin sings is derivative of the song that Redding sings, because Franklin obviously did not make it all up. I am using 'derivative' in a genetic sense rather than suggesting anything negative. Everyone agrees that, with enough change, a derivative song can be a different song than its source. Answer songs are typically seen as *different enough* that they are distinct, albeit derivative, songs in this way. This is suggested just by distin-guishing the *original song* from the *answer song*. However, one might argue that 'song' is used here to mean the track; for example, Cooper defines an answer song as 'a commercial recording' which is related to 'a previously released record' (1988: 57). More evidence that answer songs are distinct songs is provided by commercial practices. Jukebox programmers, respon-sible for buying records for jukeboxes, would typically not stock a jukebox with two versions of the same song (Billboard 1971). But one jukebox pro-grammer commented, 'I used to work in a cafe, and answer records were always played by both young and old, along with the original version of the song' (Billboard 1973b).

A further twist is provided by Stevie Wonder's 1967 cover of 'Respect' which uses Redding's lyrics. *Billboard*'s Pop Spotlight mentions the version as 'Wonder's answer song to Aretha Franklin's "Respect"— also titled the same' (1967). In the month's following Franklin's release, it had largely eclipsed Redding's original. A fairly straight rendition cover of Redding could be seen as an answer to Franklin.

In the next section, I offer a general argument that applies to answer songs and other referential covers.

Songs about songs

In earlier work, my collaborators and I offered an argument that referential covers are not instances of the same song as the original. Consider a cover which not only means something different than the original but also says something about the original version— for example, the Meatmen's cover of the Smiths' 'How Soon Is Now?' (which I discussed in Chapter 4). The Meatmen change the lyrics somewhat to provide commentary about Morrissey, the lead singer of the Smiths. Here is the argument: 'The Meatmen's cover is not merely a distinct, derivative song. It is one which is partly *about* the canonical track and the man who sings it. Its semantic content partly refers to The Smiths' track' (Magnus et al. 2013: 367). Putting this in abstract terms yields something like this:

The Songs-About-Songs Argument

1. In S's original recording of H, they are *not* singing a song that is about their version of the song.

2. In M's cover of H, they are singing a song about S's version.

3. A song which is not about S's track and a song which is about it are different songs.

4. S's original recording and M's cover are of different songs.

This is a valid argument. For the obvious substitutions of the Smiths for S, the Meatmen for M, and 'How Soon Is Now?' for H, the conclusion is that the Meatmen's cover is an instance of a different (albeit derivative) song. (The argument also works with Redding for S, Franklin for M, and 'Respect' for H.)

However, one might object to premise 2 in the argument. A version of a song can mean something that the song itself does not mean. So, even though the Meatmen are singing about the Smith's track, they might be using the same song to express something different than the original version. The reference to the Smith's original might be part of the content of the Meatmen's *performance* rather than part of the content of the *song* itself.

As an example of the difference, consider Marilyn Monroe's famous performance of 'Happy Birthday' in 1962, sung to President John F. Kennedy. The televised performance was famously described as 'making love to the president in direct view of forty million people' (Goodwin 2012). Monroe ended up meaning something very different by that performance than a roomful of first-graders means when they sing 'Happy Birthday' to a classmate. Nevertheless, Monroe and the first-graders are singing the same song.

With the difference between the meaning of a song and the meaning of a version in mind, it could be that the Meatmen take the Smiths' song and use it to say something about Morrissey. And perhaps— as Kania suggests— Aretha Franklin 'takes the content of Redding's [song] and uses it to communicate a radically different message' (2020: 242).

Kania makes a similar point by considering Cake's 1996 cover of Gloria Gaynor's 1978 hit 'I Will Survive' (2020: 241–242). Gaynor's upbeat disco track can serve as an anthem for survivors of all sorts, but Cake's cover has a disaffected, almost apathetic tone. Kania calls it a 'brilliant reimagining' (2020: 249, fn. 35). Paul Pearson writes, 'Cake tread the fine line between parody and reframing... with machine-gun guitar lines and a stand-alone trumpet that remains hilariously true to the original melody' (Treble 2018). Despite the different meaning and significance of the two versions, it seems plausible to think of them as instances of the same song.

The examples which seem most clearly to be the same song (Monroe's 'Happy Birthday', Cake's 'I Will Survive') are cases where none of the lyrics are changed. Franklin and the Meatmen mean something different than earlier versions partly by singing different words. This is suggestive but not decisive, though, because cover versions can have different lyrics than earlier versions without thereby being transformative. To recall an example from Chapter 3, consider Willie Nelson singing Paul Simon's 'Graceland.' Where Simon's lyric is 'a girl in New York City', Nelson sings 'a girl in Austin, Texas'— but nobody suggests that this substitution makes it a different song. What I suggested about that case was that the variations of 'New York City' and 'Austin, Texas' both fit within the overall meaning of the song. The lyrical changes which Franklin, the Meatmen, and others introduce in their referential covers are more substantive and— it seems to me— do not fit within the overall meaning of their original songs.

Concluding his discussion of whether covers can be instances of different songs than the original, Kania writes that 'answering this question will depend mostly on how knowledgeable rock artists and audiences treat songs and recordings of them' (2020: 242). Perhaps, but I think that drawing precise boundaries around songs will also be a matter of explication. As Bartel's survey results suggest, audiences are not terribly precise about it. We have some latitude and can be guided not just by what people already say but by what purposes we would like these concepts to serve.

The Fitzgerald dilemma

I have argued that the kind of changes which Aretha Franklin makes to 'Respect' are enough for her to be singing a different song than Otis Redding sang. This is, partly, because I think that the Songs-About-Songs Argument is sound when applied to covers which change the lyrics so as to refer to earlier versions. One might worry that, in some cases, this will lead to an awkward dilemma.

Consider Ella Fitzgerald's cover of 'Mack the Knife' (which I discussed in Chapter 4). She skips some of the material from earlier versions, and adds a verse about the fact that she is covering the song. For example, she sings in one performance, 'Bobby Darin and Louis Armstrong. They made a record (ooo, what a record) of this song. And now Ella, Ella and her fellas, we're making a wreck (what a wreck, such a wreck) of this same old song.'

The Songs-About-Songs Argument applies to Fitzgerald's cover of 'Mack the Knife', and the conclusion is that Fitzgerald *is not* singing the same song that Bobby Darin and Louis Armstrong were singing. However, Fitzgerald refers to the song that Darin and Armstrong were singing as 'this song' and says that she, too, is singing 'this same old song'— from which it follows that Fitzgerald *is* singing the same song that Bobby Darin and Louis Armstrong were singing. It cannot be both the same song and not the same song, so *either* the Songs-About-Songs Argument leads us astray here (because it is the same song) *or* Fitzgerald is saying something false (because it is not).

I can think of ways to argue that the lyrics she sings are not strictly *false*, but the details turn on the semantics of indexicals like 'this.' Even if Fitzgerald's claim using the phrase 'this song' is not strictly false, it still is not quite right. So I am tempted to accept the second horn of the dilemma: She is singing a different (albeit derivative) song.

You may not feel that this is entirely satisfactory. It may seem like there is another sense in which she is singing the same song, even if there is a sense in which she is singing a different song. The easiest way to understand such ambivalence is to think that questions about which versions are the same song do not have absolute, univocal answers.

Jeanette Bicknell recommends what she calls 'a pragmatic approach' in cases like this. When asking whether two performances are of the same song, she writes, 'we should first ask, "Who wants to know, and why?"' A singer, a musicologist, an historian, or an intellectual property lawyer might be asking for different reasons, and— Bicknell suggests— 'When people have different reasons for asking, it is not surprising if they all come up with different answers. Yet each may have good arguments for answering the question in a particular way, depending on their reasons for seeking to differentiate one song from another' (2015: 8).

This kind of pragmatic pluralism is ultimately correct, I think, but more needs to be said. It does not, by itself, provide any guidance in telling songs apart. Saying that it depends on who I am and why I care does not tell me how to get from my identity and interests to an answer. The next chapter approaches these questions in a more systematic way.

6. How a Song is Like Ducks

The last chapter stumbled over questions of song individuation. It considered cases in which it was hard to say whether two versions were— or were not— instances of the same song. In this chapter, I advocate a general metaphysical approach which helps in thinking about those cases.

The core idea is that a song is a historical individual in the same sense that a biological species is. Where a species is a lineage of organisms, a song is a lineage of versions. This supports a principled pluralism about songs.

In addition to helping with the puzzles posed in the last chapter, the view of songs as version lineages helps in thinking about mash-ups, medleys, parodies, and instrumental covers.

Songs are like species

Artworks, as Guy Rohrbaugh argues, can be understood as historical individuals. He highlights three features of artworks which we can understand by thinking of them that way. First, artworks could have had different properties than they actually do have; in technical terms, they are *modally flexible*. Second, artworks can change over time; they are *temporally flexible*. Third, artworks come into existence and could go out of existence; they are *temporal*.

Regardless of what might be said of other artworks, songs have these features. Consider the example of Bob Dylan's 'All Along the Watchtower.' It is modally flexible because, if Bob Dylan had chosen different lyrics when originally writing it, then it would have different features than it actually does. It is temporally flexible because Jimi Hendrix's cover was not something already present in the song. His innovative rendition changed the song, as revealed in subsequent versions which follow Hendrix's musical choices rather than Dylan's original. It is also temporal, because it did not exist until Dylan wrote it, and if it were forgotten and all records of it de-

https://doi.org/10.11647/OBP.0293.06

stroyed then the song would not exist anymore.

Julian Dodd objects that historical individuals are metaphysically obscure 'cross-categorial entities' and argues it would be better to deny that artworks have these features than to accept an ontology that includes historical individuals (2007: 145). In earlier work, I responded to that worry by noting that historical individuals are much-discussed in the philosophy of biology, where a standard view holds that a biological species is a historical individual. The view, originally championed by Michael Ghiselin (1966, 1974) and David Hull (1976, 1978), is now widely held. Although it would be an overstatement to call it a consensus, pockets of opposition are motivated by the details of biology rather than by the thought that historical individuals are somehow incoherent. What is respectable for science is respectable for art. I argue elsewhere that instrumental musical works are like species (Magnus 2013), and a similar argument is given by Charles O. Nussbaum (2007,2021). Here I want to think of songs as historical individuals in much the same way that species are.

Imagine you go to a pond and see a flock of common mallards out on the water. Focus on a particular middle-aged mother duck. She is a historical individual who started as a duckling, developed and matured, mated. She will, if she is lucky, grow to old age. Eventually, she will die.

That single duck is part of the species, *Anas platyrhynchos*. She is one branch of a lineage that goes back through her parents to generations of ancestors and (with any luck) will go forward through her offspring to generations of descendants. Each of the mallards on the pond is part of that same species, connected to each other and to all other mallards by natural processes of reproduction and development. Thinking of the species as a historical individual means thinking of each organism which is a member of the species as part of that lineage.

A song like 'Happy Birthday' can be sung in different places and at different times. Each performance can be thought of as an individual in the same way that each separate mallard can be. Yet all the performances of 'Happy Birthday' are connected by cultural and musical processes. To think of them as performances of the same song is to think of them as part of that lineage, which is a larger historical individual in the same way that the species *A. platyrhynchos* is.

I started with ducks because they are easy to visualize and because they are funny, but songs are different than ducks in one important respect. Almost all ducks result from sexual reproduction, meaning that each offspring varies from its parents. Performances of songs are like that. Even if the same performer endeavors to perform a song in the same way multiple times, each performance is different. However, songs can also be recorded. A recording can be played back multiple times. If I listen to a digital track and then play it again, what comes out of my speaker is the same track twice. So a better analogy is with a plant species.

Imagine, some distance away from the ducks, a patch of woodland strawberries. The species, *Fragaria vesca*, propagates in two ways. It grows fruit with seeds, which can sprout and grow into new plants. What it does mostly, though, is send out runners: horizontal stems which grow a distance out from the plant, tuck into the ground, and start growing a separate root system and leaves. When the runner withers away, the separate root system can sustain a separate plant. Propagation by runners produces clones of the original plant. This means that the whole patch of woodland strawberries, if propagated entirely by runners, might have all the same genome. We can see each plant in the patch as an individual, but there is also a sense in which the whole patch is the same individual.

There is a new version of a song every time it is performed, just as there is a new organism every time a strawberry seed sprouts. There is also a new version of the song when a new track is mastered in the studio. When that track is recorded as a sound file, sent over the internet to my computer, and I play it back, the process is like a runner propagating a new strawberry plant. The sounds coming out of my speakers are more a clone than a new version. The track altogether, including all of the times it is played, is part of the song as a historical individual.

One might point out that the ducks on the pond and the strawberries in the field have done their natural thing, whereas songs are the work of people. That does not show that one is not a historical individual, though, just that the causal processes which hold one kind of individual together are different than the causal processes that hold another together.

Developing the analogy between songs and species further is the task of the rest of this chapter. It will ultimately help in finding solutions to the puzzles about song individuation which I raised in Chapter 5 as well as to some other puzzles about covers.

Before moving on, I want to point out the metaphysical modesty of my position. Ghiselin claims that the view of species as individuals 'provides the inspiration for a new ontology with profound implications for knowledge in general' (2009: 254). Rohrbaugh himself describes his approach as 'innovation at the level of metaphysics, the identification of a new ontological category' (2003: 197). However, I am not claiming that historical individuals are among the fundamental building blocks of reality. A historical individual might be thought of as a composite thing or, as John Dupré suggests, a process (2021). Adapting the approach of Richard Boyd, I suggested in earlier work that a historical individual is a particular kind of *Homeostatic Property Cluster* (HPC) (Boyd 1999, Magnus 2013). Nussbaum instead adopts Ruth Garrett Millikan's far-reaching metaphysical picture and sees a historical individual as a *Reproductively Established Family* (REF) (Millikan 1984, Nussbaum 2007). It has also been suggested that a historical individual might be an abstract object which ontologically depends on but is not constituted by its embodiments (for which there does not seem to be an acronym). Interested readers are welcome to look elsewhere for details of those debates. Any of these proposals would suffice for my purposes here.

Species pluralism

In order to develop the analogy between songs and species further, let's briefly consider some philosophy of biology. (I have written about this at greater length elsewhere (Magnus 2012: 83–96).) Minimally, a species is a lineage: a series of organisms connected by relations of descent. Trying to specify which lineages are species faces pluralism in two respects.

First, *rank pluralism*: It is often indeterminate as to whether a particular group should count as a species, a less specific rank such as a genus, or a more specific rank such as a subspecies. This determination is guided by 'real biological attributes' but is nevertheless 'semisubjective' (Baum 2009: 76). Rank pluralism can be especially vexing in the case of incipient species, a subgroup within a species which is somewhat isolated and on a trajectory to become a distinct species.

Second, *concept pluralism*: There are distinct but legitimate features which might chart the boundary of a species. According to the biological species concept, a species is a reproductively isolated, interbreeding group— members of different species either cannot breed or produce infertile offspring. According to the ecological species concept, a species is a group of organisms that fill an adaptive zone or niche— members of different species do not typically breed, but whether they *could* is irrelevant. The details of these need not concern us, because the point of the analogy is just that there are different features which might be relevant for identifying organisms as members (or not) of the same species.

I am not sure whether every kind of historical individual will admit of rank pluralism and concept pluralism, but species do— and it seems to me that songs do, too.

My idea is to see a song as a lineage of versions connected by relations of inspiration and copying. Just as the members of a species are organisms of common descent, the versions of a song are performances/recordings with historical continuity and causal dependency. Just as biologists can make judgements about species given the interests and considerations salient in a particular case, people interested in music can make judgements about songs.

Rank pluralism for songs

Rank pluralism can be illustrated with several examples which came up in earlier chapters.

First: When asked to change the word 'cocaine' to 'champagne' in his song 'Listen to Her Heart', Tom Petty refused on the grounds that 'it would have made it a different song' (Zollo 2012: ch. 16). Yet imagine a cover of 'Listen to Her Heart' which followed the music and lyrics of the original and changed just that one word. I suspect that many fans would consider it the same song. Petty's insistence to the contrary should make one wonder whether it is a larger change than one would otherwise imagine. Even so, Petty is not obviously right. There is no precise amount of difference that makes two versions different songs.

Second: I argued in Chapter 5 that Aretha Franklin's cover of Otis Redding's 'Respect' is a different song than Redding's. Franklin's version has become the target of countless covers, and someone might cover it without knowing about Redding's version. Women who sing 'Respect' as a showcase for their vocal prowess often perform fairly straight rendition covers of Franklin's version. Kania suggests that my coauthors and I 'are misled into thinking that Franklin's "Respect" is a recording of a different song than Redding's because her version has become the standard that later covers take as their target (from which most people learn the song, etc.)' (2020: 242). Contra Kania, the fact that Franklin's version became canonical makes an important difference, and I do not think that we are misled at all. In the biological analogy, Franklin's cover was the first member of an incipient species. If it had no descendants, then it might just as well be counted as a mutant version in the species of Redding's original. But it was fruitful, the first member of a new lineage.

Rank pluralism about songs means that this judgement (that Franklin's version is a different song than Redding's) is somewhat subjective. However, it responds to real musical and historical features of the case. Franklin's version does start a separate lineage, and covers of Franklin's version are a distinct kind of version from direct covers of Redding's. The element of subjectivity is whether we call it a separate *song* as opposed to a *sub-song* (or whatever we want to call the song analog of a sub-species).

Third: I suggested in Chapter 5 that Hilary Duff's 2008 'Reach Out' is an interpolated cover. That is Claire McCleish's terms for a cover that uses the chorus or refrain from an earlier recording to connect new lyrics and rapped verses (2020: ch. 5). Whereas McCleish sees an interpolated cover as almost but not quite the same song as the original, Dan Burkett suggests that 'Reach Out' is simply the same song as Depeche Mode's 'Personal Jesus' (2015). Admittedly, a cover of 'Reach Out' would sound strikingly different than a straight rendition cover of 'Personal Jesus.' Nevertheless, Duff's track is descended from Depeche Mode's. They form part of a lineage. The subjectivity is just in deciding whether we call them the same song.

These examples show that questions of whether two versions are the same song will not always have univocal or straightforward answers. Rank pluralism means that different answers are possible. Nevertheless, any answer must respond to the underlying musical and historical features of the versions.

Concept pluralism for songs

Concept pluralism requires that there be different ways of thinking about what makes two versions the *same song*. Let's start with the distinction between song identity defined by lyrical content and identity defined by musical features. By highlighting lyrical continuity, we can recognize genre-shifted covers that change the sound of the song considerably. By highlighting musical features, we can recognize instrumental covers that omit lyrics entirely.

Note that this resolves the Fitzgerald dilemma, from the end of the last chapter. Considered in terms of musical features, Ella Fitzgerald's 'Mack the Knife' is straightforwardly the same song as Louis Armstrong's and Bobby Darrin's versions. Her improvised lyrics follow the melody of the original. Considered as words, though, the new lyrics make for a different song. The space created by pluralism makes it possible to recognize both intuitions as legitimate. The underlying fact is that her version is in the lineage of earlier versions. One might consider it as just a remarkable mutant in that line; if other people started covering Fitzgerald's version, one might instead see it as the progenitor of a new line.

Let's consider three examples in more detail.

First, consider the case of George Harrison's 1970 'My Sweet Lord.' The musical structure strongly resembles that of the 1963 hit 'He's So Fine', written by Ronnie Mack and made famous by the Chiffons— so much so that the publisher which owned the rights to 'He's So Fine' sued for infringement. In his ruling on the case, Judge Richard Owen writes, 'It is clear that "My Sweet Lord" is the very same song as "He's So Fine." This is, under the law infringement of copyright and is no less so even though subconsciously accomplished' (NY Times 1976). Of course, the law does not always get ontology right. What counts as the same for the law might not be the same in fact. Note, however, that the legal question was whether Harrison's track counted as copyright infringement. There was not and is not a legal sense to 'same song', so the judge can be understood as applying— as well as he can— the ordinary sense of the words. By focussing on musical features, he judges the two to be the same song.

The lyrics to 'My Sweet Lord' are about religious transcendence. The background vocals juxtapose 'Hallelujah' with 'Hare Krishna', 'Hare Rama', 'Guru Brahma', and other Hindu mantras. In contrast, the lyrics to 'He's So Fine' are about romantic attraction. The nonsense syllables 'Do-lang-do-lang' are repeated in the background vocals. Although religious transcendence can be a metaphor for romantic love (and vice versa) both 'My Sweet Lord' and 'He's So Fine' strike me as being literal about their topics. The mantras in Harrison's lyrics do not mean the same thing as the nonsense syllables in Mack's lyrics. So, focussing on lyrics, the two are obviously different songs.

The point could be put this way: Someone playing the instrumental part of 'My Sweet Lord' is also playing 'He's So Fine.' Someone singing 'My Sweet Lord' is not also singing 'He's So Fine.' Highlighting the former, *same song*; highlighting the latter, *different song*.

Second, consider Paul Anka's song 'My Way.' It is set to the tune of Claude François' 'Comme d'habitude', but Anka did not translate the French lyrics. In fact, he says that he thought the French original was 'a shitty record, but there was something in it' (McCormick 2007). Anka acquired publishing rights to the song and, after a conversation with Frank Sinatra, wrote new lyrics from Sinatra's point of view. If we focus on lyrics, 'My way' and 'Comme d'habitude' are obviously different songs— and that is how most people think about them. Why? Perhaps just because Anka secured legal rights to the tune. In the case of 'My Sweet Lord', the legal question of copyright infringement foregrounded the sense in which it was the same as 'He's So Fine.' Talk about 'My Way' tends to focus instead on the meaning of the lyrics, as when contrasting Sinatra's original version with Sid Vicious' cover. Because Anka had made a deal for the rights, the continuity with 'Comme d'habitude' is not especially interesting.

If pluralism is correct, though, there ought to be at least some contexts in which 'My Way' and 'Comme d'habitude' are described as the same song— and there are. Jacques Revaux originally wrote the tune without lyrics and sent them to a publisher in London who would add lyrics and record a demo. English language lyrics were written by a young David Bowie, as 'For Me', and a demo was offered to several artists with no success. Revaux later ended up playing the demo for François, who wanted to use lyrics that he had written independently. A website featuring an interview with Revaux treats the demo of 'For Me', François' 'Comme d'habitude', and Anka's 'My Way' as the same song (What The France 2020). That makes sense, because talking to the songwriter who wrote the melody and bridge of the song foregrounds musical features.

In this case, too: Someone playing the instrumental part of 'My Way' is also playing 'Comme d'habitude.' There are instrumental versions posted on-line labelled as being covers of both. However, someone singing 'My Way' is not singing 'Comme d'habitude.' Highlighting the former, *same song*; highlighting the latter, *different song*.

The examples so far were ones where a focus on musical features supported a judgement of *same song* but a focus on lyrics supported a judgement of *different*. The reverse is also possible.

Third, consider 'House of the Rising Sun', a traditional song probably best known in the Animals' 1964 hit version. Jeanette Bicknell describes a friend who thinks that the song *must* be in a minor key and have triple meter (like the Animals' version). Confronted with Woody Guthrie's 1941 version, which is in a major key and has duple meter, her friend protests that it is not 'House of the Rising Sun' at all. 'The words are the same,' she says, 'but changed harmonic structure, melody, time signature, key. There's a breaking point and I can't define it but I know it when I hear it. This is Woody Guthrie's song "X", with words from "House of the Rising Sun"' (2015: 15). Bicknell herself disagrees, arguing that Guthrie's song is a *bona fide* instance of 'House of the Rising Sun' precisely because it shares lyrics with other versions of the song. She writes, 'Treating these major and minor key versions as being of "the same" song allows us to ask questions about the song's trajectory across time and the ways in which different communities have changed it' (2015: 16). It seems to me that, recognizing the fact of pluralism, both Bicknell and her friend are right. Bicknell appeals to lyrical continuity so as to trace a larger historical individual, while her friend appeals to musical difference for aesthetic and appreciative reasons.

One might object that neither party should accept this rapprochement. According to the objection, pluralism is just a vacuous attempt to let everybody be right.

The objection fails for several reasons. First, even given pluralism, same-song judgements must still be responsive to facts about the versions in question. In order for two versions to be of the same song in even one respect, there must be a sufficient historical connection between them and they must share features upon which that respect depends. If we are interested in the lineage that includes both Guthrie's version and the Animals', we might treat them as members of the same genus rather than as the same species— there is a lineage that includes both, regardless of how we parse that lineage into songs. Second, pluralism provides an explanation of conflicting or diverging intuitions in particular cases. Bicknell and her friend reach different conclusions precisely because the difference arises from employing different song concepts. Third, it is productive to think in terms of version-lineages rather than demanding univocal answers to questions about song identity. All the cases I discuss in this chapter are meant to serve as examples of that.

Lyrics as words, lyrics as meaning

Making same-song judgements on the basis of lyrical content may look at the specific words used, as in the examples which I discussed in the previous section. It is possible instead to look at the function of the lyrics— their contribution to the overall meaning of the song.

To recall a point from Chapter 3: As Theodore Gracyk notes, songwriters may have 'open-ended intentions' such that 'elements of a lyric that seem to refer to concrete things and situations are merely exemplary' (2001: 66). Willie Nelson sings about 'a girl in Austin, Texas' instead of 'a girl in New York city' when singing Paul Simon's 'Graceland', and that substitution fits with the core meaning of the line and the song. So Nelson's version counts as the same song even though he sings different words.

In other cases, preserving the function and meaning *requires* that the lyrics change between versions. Bicknell gives the example of 'Happy Birthday', which changes on each occasion so that it is addressed to whoever is having a birthday (2015: 8).

So lyrical faithfulness can be understood either as singing the same words or as singing words that preserve the overall meaning. This difference explains what was vexing about the covers of 'California Über Alles', discussed in the previous chapter. The cover by Vio-Lence uses the same lyrics as the original, singing about Jerry Brown and 'Carter power' in 2020. It is the same song as the original (considered with respect to preserving exact words) but a different song (considered with respect to meaning). It is a bit like singing 'Happy birthday to Greg' on every occasion, because there was a guy named Greg who once had a cool birthday party. The Disposable Heroes Of Hiphoprisy's cover changes the lyrics. There is no absolute standard for how much difference is required for it to count as a different song, but perhaps the Disposable Heroes' cover is a different song considered with respect to preserving exact words. If we see the song instead as a specified meaning, a structure for commenting on the current California governor, then the Disposable Heroes' lyrical changes are faithful to the original.

A further example is provided by JK-47's 2021 cover of Tupac Shakur's 1998 track 'Changes.' (The original 'Changes' itself is an interpolated cover, sampling the hook of Bruce Hornsby's 1986 'The Way It Is.' In JK-47's cover, the hook is sung by Bronte Eve rather than being sampled.) Jacob Paulson, who performs under the name JK-47, is an indigenous Australian. His cover begins and ends with a sentence in the language of his people, and it leaves out Tupac's lyrics which refer specifically to the American situation. Yet Paulson retains the lyrics from the original which he felt were the most powerful. He says, '[Tupac] was talking about something real, and that's what I wanted to do. Tupac goes in… for how it is in America, so I had to keep it real and tell… how it is in Australia and how it is as an indigenous person' (JJJ 2021a). By focussing on the specific words that Tupac raps in his original, JK-47's cover might count as a different song. By focussing on the core meaning of the song, however, it is plausible that JK-47 would have been doing something different if he had rapped the exact same words. To fit the core meaning of the song, he had to rap about his own situation.

Concept pluralism requires that there be multiple ways of thinking about what makes versions the *same song*, and I have argued for three: musical features, lyrics (considered as specific words), and lyrics (considered as overall function or meaning).

Further puzzles resolved

In the remainder of the chapter, I want to consider some other puzzle cases which can be better understood by thinking of songs as version lineages: mash-ups and medleys, parodies, and instrumental covers.

Covers, mash-ups, and medleys

A cover can target more than one earlier version. In the simplest case, a cover can target a version which is itself a cover such that both the earlier cover and the original are important influences. In Chapter 4, I gave the example of college a cappella groups covering 'Bitches Ain't Shit.' The groups take the lyrics from Dr. Dre's original and musical features (including the stark genre shift) from Ben Folds' earlier cover. Folds' cover is the direct offspring of Dre's original. Because the a cappella group is informed by both the original and the earlier cover, it is the offspring of both of them— not something that occurs in biological lineages, but something which can easily occur in musical version lineages. The complete lyrical continuity makes it plausible to see them all as versions of the same song.

More puzzling are versions which cover earlier versions from two separate lineages. In live performances in the mid-2010s, Chris Cornell took the music from U2's 'One' (from their 1991 album *Achtung Baby*) and the lyrics from Metallica's 'One' (a different song, from their 1988 album *And Justice For All*). Cornell explained that he searched the internet for U2's lyrics but that Metallica's lyrics had come up as the first hit. Musician and producer Rick Beato calls it 'one of the most creative cover songs of all time' (2019). Is it really a cover? Blogger Steven Richard refers to it instead as a 'mash-up' (2018). I am unsure whether it is right to call it a mash-up either, because mash-ups in the usual sense digitally combine elements from two tracks into one recording. For example, the Reddit community *r/mashups* defines a 'mashup' as 'a song or composition created by blending two or more pre-recorded songs, usually by overlaying the vocal track of one song seamlessly over the instrumental track of another.' Regardless, the ontology of Cornell's 'One' is similar to that of a mash-up. As Christopher Bartel argues, 'Mash-ups are musical works in their own right, and yet also happen to be interesting cases of works that instantiate parts of other works' (2015: 305). Cornell sings a song— a work in its own right— made from parts of two other songs. At the cost of introducing jargon, we might call it a *mash-up cover*. (The digital kind of mash-up would then be a *mash-up remix*.)

As another example, consider Jon Sudano's YouTube videos which he labels as covers. In dozens of videos posted from 2016 to 2019, Sudano plays the music from the track he claims to be covering but sings the lyrics of Smash Mouth's 'All Star.' After getting through a chorus of 'All Star', he sings a few words from the ostensible target of his cover. Like Cornell's dual version of 'One', Sudano's versions set the lyrics of one song to the tune of another. One reporter describes them both as 'covers' and as 'Smash Mouth mashups' (Siese 2016). That is, they are mash-up covers.

Mash-up covers differ from typical covers in the way that a hybrid differs from a typical species. They are a version resulting from two distinct lineages. If other artists were to cover Cornell's 'One' or one of Sudano's All-Star performances, then it would start new lineage and a novel song. If that does not happen, then there is little need to make same-song judgements and it can be regarded as a hybrid oddity.

Note that the causal influence for mash-up covers is different than for the more familiar case of medleys. For example, when asked to perform a cover of the Cyndi Lauper song 'She Bop', the band Gwar begins with a genre-shifted rendition and then 'seamlessly transitions into an excellent cover of the Ramones' "Blitzkrieg Bop"' (Kurp 2015). It is a medley, with a clear distinction between the first section (when they are covering Lauper) and the second section (when they are covering the Ramones). More complicated medleys work similarly. For example, Weird Al Yankovic's many polka medleys stitch together several different songs, set to accordian and sung in Yankovic's distinctive voice. Ray Padgett notes that 'these polka-fied medleys of popular songs were honest-to-goodness covers. He sang the words straight, not adding any of his own lyrical jokes. The humor came in the music, recontextualizing well-known songs into ridiculous... arrangements' (2017: 142). Even though more songs are melded together in Yankovic's medleys, the sections still appear one after another.

Although a medley and a mash-up both instantiate parts of other songs, the difference is that a medley only instantiates one at a time. The parts of a medley are played in series, first one and then the other. In contrast, a mash-up includes parts of both source songs simultaneously. In his version of 'One', Cornell plays parts of U2's song and Metallica's song in parallel.

This underscores the fact that identifying a version's pedigree— saying what lineage it is in— is not just a matter of identifying its influences. It also requires recognizing what parts of the earlier versions appear in the new one and in what combination. The details of the causal influence matter.

Parodies

Aside from the medleys, Weird Al Yankovic mostly makes parodies. For example, 'Eat It' has lyrics about eating food set to the tune of Michael Jackson's 'Beat It.' These are typically not considered covers even though, as I noted in Chapter 1, it can be hard to say why. Although his earliest parodies had accordion parts in rock and pop songs, his later parodies have often been crafted so as to sound like the originals. Yankovic and his long-time band labor over that effect, so that a listener hearing a parody out of context might mistake it for the original. Yankovic says, in an interview with Lily Hirsch, 'It's like a forensic kind of thing where we... try to figure out everything that the original musicians did in the studio. And if we're baffled, sometimes we'll approach the original musicians.' Drummer Jon Schwartz describes it as 'an exercise in backwards engineering.' Hirsch also notes that the work to make the sound match goes beyond just what the musicians play and how. Production engineers are also crucial to making the parody track sound as much like the original track as possible. (See Hirsch 2020, ch. 2.)

Admittedly, the instrumental parts are not quite mimics. Yankovic's parodies are often sped up somewhat, to enhance the comedic energy. And sometimes changes are made to bring the melody within his vocal range. Even so, I have different intuitions about the question *Is Weird Al singing the same song as Michael Jackson?* than I do about the question *Is Weird Al's band playing the same song as Michael Jackson's band?* Their careful effort to make the instruments and production sound the same are enough to shift my focus to musical features, and I find *Yes* a tempting answer to the latter question.

Regardless of whether we call it a cover, a track like 'Eat It' is a child of 'Beat It' in much the same way a cover would be. When the Japanese punk band Shonen Knife covered 'Eat It', they were making a new version in a lineage that goes back to Michael Jackson.

Not all of Yankovic's parodies are like that, however. Some are original comedic songs in a particular style. For example, his original song 'Dare to be Stupid' is a parody in the style of Devo, and his 'You Don't Love Me Anymore' is a parody in the style of Extreme's 'More Than Words.' In the former case, 'Dare to be Stupid' is not the offspring of any one Devo song, nor is it descended from all Devo songs the way that a medley or mash-up would be. In the latter case, 'You Don't Love Me Anymore' is inspired by a particular song but only in its overall style. This is a kind of descent that has no clear analogy in biological lineages.

Hirsch provides some helpful vocabulary to distinguish these possibilities. Parodies which target specific earlier tracks or songs, in the way that 'Eat It' parodies 'Beat It', are *direct parodies*. Parodies which target a style or feel, in the way that 'Dare To Be Stupid' parodies Devo, are *style parodies*. Direct parodies extend a lineage in the way that covers do, whereas style parodies do not.

Even among direct parodies, it is worth distinguishing between different ways the parody can relate to its target. Parodies like 'Another One Rides the Bus' and 'Eat It' have lyrics that match the meter of the original but are about unrelated things. Although they are direct parodies, they are also *non sequiturs*.

In other cases, Yankovic writes lyrics that are about the original song. His 1992 'Smells Like Nirvana' not only echoes the musical features of Nirvana's 1991 hit 'Smells Like Teen Spirit', but also gently mocks the mumbled singing of Nirvana's Kurt Cobain. Even if it is not a cover, it refers to Cobain and Nirvana much in the way that the Meatmen's cover of 'How Soon Is Now' refers to Morrissey and the Smiths. Contrast Pansy Division's 'Smells

Like Queer Spirit', also from 1992. The Pansy Division track changes the lyrics to make it a song about growing up gay. Jon Ginoli of Pansy Division insists, 'Not a parody, an affectionate tribute' (Lapriore 2001). Regardless, the change makes it about something else entirely.

Another referential direct parody is Yankovic's 2011 'Perform This Way', a parody of Lady Gaga's 'Born This Way' from the same year. The lyrics of the parody are a send up of Lady Gaga's performance persona— but it is also a celebration of individuality and personal expression, just like the original. The difference between the two is not so much the overall message as that Yankovic's version is packed with jokes. In terms of overall meaning, 'Perform This Way' may be the same song despite being a parody.

Instrumental covers

Popular usage readily identifies instrumental versions of songs as covers. This shows that it is possible for a version to count as a cover without retaining any of the lyrics from the original. There is a perfectly ordinary sense in which one can hear an instrumental version of a pop song, played perhaps in the background at a shopping mall, and say 'I know that song!' That is, an instrumental cover can count as the same song as its lyrical original.

It is less clear what to say about the rare cases of covers adding vocals to tunes that started out as instrumental. For example, Hugh Masekala's recording of the Philemon Hou composition 'Grazing in the Grass' was a hit in 1968. The group the Friends of Distinction added lyrics and had a hit with their cover in 1969. Is the Friends' cover the same song as Masekala's original? It is tempting to say that it is, just as an instrumental version of the Friends' 'Grazing in the Grass' would be the same song. Yet it is unclear what we should say about an alternative version which added completely different lyrics to the music of the original. Would it be the same song as the Friends' version?

Fortunately, this does not need to be just a thought experiment. Consider the tune 'Popcorn', written by Gershon Kingsley and released on his 1969 album *Music To Moog By*. A 1972 version by the band Hotbutter reached #9 on the Billboard Pop Singles chart. There have been literally hundreds of covers, some of which have added lyrics. A 1972 version by Anarchic System opens with the lyrics, 'Like a pop-corn in your hand / is your castle made of sand / life goes up and life goes down / and life goes round and round and round.' Several subsequent covers have used the lyrics from that version, but other artists have written their own. A 2003 version by Fiddler's Green, making no mention of popcorn, opens 'There's a cold wind blowing. / Will you find your way home?' A 2011 version by the Brits and Pieces simply has 'You've got popcorn in your mouth' and rhythmic repetition of the word 'pop.' (Coen van der Geest's website at popcorn-song.com is a treasure trove of information.)

Are these versions of 'Popcorn' all the same song as Kingsley's original? Are they the same song as each other? With the concept of musical features in mind, the answer is *Yes*. With lyrics in mind, *No*.

Thinking in terms of lineages and pluralism: All the covers are offspring of Kingsley's version, just as The Friends' 'Grazing in the Grass' is an off-spring of Masekala's. The versions which add different lyrics to 'Popcorn' are siblings to one another. Musical and historical connections determine a family tree, but we have some freedom in parsing this family tree into songs.

Conclusion

The last chapter ended with Jeanette Bicknell's suggestion of a pragmatic approach to questions of song identity. The answer to whether two versions count as the same song, she says, should depend on 'Who wants to know, and why?' The view of songs as historical individuals fits nicely with such an approach.

The reason it matters who is asking is not because of some wild relativism where everyone gets their own reality. Rather, there are versions standing in different relations of inspiration and causation. These version lineages exist, regardless of who is asking.

Because of rank pluralism, one person might judge a certain degree of difference to be enough for two versions to be different songs while another person does not. Because of concept pluralism, people might look to different kinds of features in judging whether two versions are the same song. Given their different interests, they might privilege musical features, the words of the lyrics, or the overall meaning of the lyrics— and these can lead to different answers.

Epilogue

'Many's the time I've been mistaken
And many times confused.'
— Willie Nelson, covering Paul Simon's 'American Tune' (1993)

So where does this wide-ranging discussion of covers leave us?

In discussions about music, there are plenty of tracks and performances that we readily identify as cover versions. I have suggested that we should not try to refine the concept much further than that. Any would-be definition is subject to counterexamples, and it would be pointless logic-chopping to insist that versions must be sorted decisively into covers and non-covers.

Some people disdain covers, but that dim view is usually the result of thinking that all covers are attempts at slavish imitation. That is, it is the result of failing to distinguish between mimic covers and rendition covers.

At best, a mimic cover can be a superfluous duplicate. At worst, it can be a terrible mess. A rendition cover, however, can succeed or fail in a variety of ways. A straight rendition cover might be bad for being too much like the original, or it might reflect the canny choice to keep elements that work well. A wildly transformative rendition cover might be bad or good, depending on how the changes play out.

A rendition cover can differ from the original in musical elements, in words, or in meaning. Some covers use the difference to refer to other versions, like answer songs that are in dialogue with the lyrics of the original. It is plausible to think that these changes are enough to make them different songs. Ultimately, though, I recommended pragmatic pluralism: In judging versions to be the same or different songs, we can evaluate similarity in different degrees and respects. Regardless of whether we call them the same song or not, the original and the cover stand in a relation of inspiration. They are part of the same historical individual, a trajectory of music making.

 https://doi.org/10.11647/OBP.0293.07

A rendition cover can be appreciated with that trajectory in mind, in relation to the original, or just as its own thing. I have resisted the suggestion that there is some general rule which can tell us which way of appreciating a cover is proper or best. There is no substitute for listening and responding to the versions in question.

We are left, in the end, listening to music.

References

The URLs given are current as of the writing of this book, except where noted.

Aesthetics for Birds. 2014. 'August is covers month at AFB', 1 August <https://aestheticsforbirds.com/2014/08/01/august-is-covers-month-at-afb/> 'AFB covers contest winners!', 6 September <https://aestheticsforbirds.com/2014/09/06/afb-covers-contest-winners/>

Appel, Nadav. 2018. 'Pat Boone's last laugh: Cover versions and the performance of knowledge', *International Journal of Cultural Studies*, 21(4): 440–454.

Apple Music style guide. 2021. version 2.1.7, August <https://help.apple.com/itc/musicstyleguide/en.1proj/static.html>

Bartel, Christopher. 2015. 'The metaphysics of mash-ups', *Journal of Aesthetics and Art Criticism*, 73(3): 297–308

——. 2017. 'Rock as a three-value tradition', *Journal of Aesthetics and Art Criticism*, 75(2): 143–154

——. 2018. 'The ontology of musical works and the role of intuitions: An experimental study', *European Journal of Philosophy*, 26(1): 348–367

Baum, David A. 2009. 'Species as ranked taxa', *Systematic Biology*, 58(1): 74–86

Beato, Rick. 2019. 'Top 10 cover songs of all-time', 30 April <https://www.youtube.com/watch?v=wcXgv0SCWqQ>

——. 2021. 'Reacting to the top 10 songs on iTunes. . .BTS WTF?', 26 May <https://www.youtube.com/watch?v=XG9Y2LBuoww>

Bicknell, Jeanette. 2015. *Philosophy of Song and Singing: An Introduction* (New York: Routledge)

Billboard. 1955a. 'There ought to be a law: Lavern Baker seeks bill to halt arrangement "thefts"', 5 March, pp. 13, 18

——. 1955b. 'WINS issues ban on copy records', 27 August, p. 21

© P.D. Magnus, CC BY 4.0   https://doi.org/10.11647/OBP.0293.08

——. 1967. 'Pop spotlight: I was made to love her', 30 September, p. 82

——. 1968. 'U.S.' Spanish chart inroads, fall of cover are cited by Milhaud', 19 October, p. 75

——. 1970. 'Cover record waste —says Janus chief', 11 July, p. 4

——. 1971. 'Programmers in the news', 9 January, p. 38

——. 1973a. 'The evolution of children's records: Rising manufacturing costs endanger low price kiddie disks', 7 July, p. 40

——. 1973b. '"Tie a ribbon" answer poses cover query— programmers waiting', 7 July, p. 29

——. 1974. 'Billboard's top single picks', 29 June, p. 78

Blush, Steven. 2010. *American Hardcore: A Tribal History*, second edition (Feral House)

Boyd, Richard. 1999. 'Homeostasis, species, and higher taxa', in *Species: New Interdisciplinary Essays*, ed. by Robert A. Wilson (Cambridge, Massachusetts: MIT Press), pp. 141–185

Brown, Lee B. 2014. 'A critique of Michael Rings on covers', *Journal of Aesthetics and Art Criticism*, 72(2): 193–195

Brown, Sandy. 1968. 'Discovering Tiny Tim', *The Listener*, 7 November, pp. 622–623

Bruno, Franklin. 2013. 'A case for song: Against an (exclusively) recording centered ontology of rock', *Journal of Aesthetics and Art Criticism*, 71(1): 65–74

Bundy, June. 1961a. '"Oldie" programming move grows', *Billboard*, 23 January, pp. 1, 47

——. 1961b. 'DJ's toast oldie trend', *Billboard*, 6 February, pp. 1, 41

——. 1961c. 'Vintage disks stealing show', *Billboard*, 17 April, pp. 1, 8

Burkett, Dan. 2015. 'One song, many works: A pluralist ontology of rock', *Contemporary Aesthetics*, 13(13) <http://hdl.handle.net/2027/spo.7523862.0013.013>

Cappelen, Herman. 2018. *Fixing Language: An Essay on Conceptual Engineering* (Oxford University Press)

Carnap, Rudolf. 1962. *Logical Foundations of Probability*, second edition (Chicago: University of Chicago Press)

——. 1963. in *The Philosophy of Rudolf Carnap*, ed. by Paul Arthur Schilpp, volume XI of Library of Living Philosophers (La Salle, Illinois: Open Court)

Cashbox. 1957. 'A new concept of the music business', 25 May, p. 95

Cooper, B. Lee. 1988. 'Bear Cats, Chipmunks, and slip-in mules: The "answer song"; in contemporary American recordings, 1950–1985', *Popular Music and Society*, 12(3): 57–77

Coplan, Chris. 2015 'Update: Tom Petty awarded songwriting royalties for Sam Smith's "Stay With Me"', *Consequence*, 29 January <https://consequence.net/2015/01/tom-petty-awarded-songwriting-royalties-for-sam-smiths-stay-with-me/>

Covach, John, and Andrew Flory. 2018. *What's that sound?*, fifth edition (New York and London: W. W. Norton & Company)

Coyle, Michael. 2002. 'Hijacked hits and antic authenticity: Cover songs, race, and postwar marketing', in *Rock Over the Edge: Transformations in Popular Music Culture*, ed. by Roger Beebe, Denise Fulbrook, and Ben Saunders, editors (Durham: Duke University Press), pp. 133–157

Cummings, Jon. 1994. 'Shawn Colvin models others' songs', *Billboard*, 9 July, pp. 14, 18

Cusic, Don. 2005. 'In defense of cover songs', *Popular Music and Society*, 28(2): 171–177

Davies, Stephen. 2001. *Musical Works and Performances: A Philosophical Exploration* (Oxford: Clarendon Press)

DiGangi, Diana. 2019. ''80s pop star Tiffany reboots nostalgic smash hit "I Think We're Alone Now"', *WLJA*, 1 May <https://wjla.com/news/entertainment/80s-pop-star-tiffany-reboots-nostalgic-smash-hit-i-think-were-alone-now>

Dodd, Julian. 2007. *Works of Music: An Essay in Ontology* (Oxford University Press)

Dupré, John. 2021. *The Metaphysics of Biology*, Elements in the Philosophy of Biology (Cambridge University Press)

Easy Song Team. 2021. 'What is an "interpolation"?' <https://support.easysong.com/hc/en-us/articles/360038912034-What-is-an-Interpolation->

Edwards, Bob. 2002. 'Johnny Cash, the Man in Black's musical journey continues', 6 November <https://www.npr.org/2002/11/06/833464/johnny-cash-the-man-in-blacks-musical-journey-continues>

Frith, Simon. 1996. *Performing rites: On the value of popular music* (Cambridge, Massachusetts: Harvard University Press)

Gabler, Milt. 1955. 'Talent and tunes', *Cashbox*, 27 July, p. 2

Gezari, Janet, and Charles Hartman. 2010. 'Dylan's covers', *Southwest Re-*

view, 95(1/2): 152–166

Ghiselin, Michael T. 1966. 'On psychologism in the logic of taxonomic controversies', *Systematic Zoology*, 15(3): 207–215

——. 1974. 'A radical solution to the species problem', *Systematic Zoology*, 23(4): 536–544

——. 2009. Metaphysics and classification: Update and overview, *Biological Theory*, 4(3): 253–259

Glamour. 2018a. 'Halsey watches fan covers on YouTube — Glamour', *YouTube*, November 30 <https://www.youtube.com/watch?v=15-kO5sAOy4>

——. 2018b. 'Hayley Kiyoko watches fan covers on YouTube — Glamour, *YouTube*, December 27 <https://www.youtube.com/watch?v=O8F6a kdgvZQ>

Gleason, Ralph J. 1973. 'Perspectives: "cover" versions and their origins', *Rolling Stone*, 7 June, p. 7

Goehr, Lydia. 2007. *The Imaginary Museum of Musical Works*, revised edition (Oxford University Press)

Goldblatt, David. 2013. 'Nonsense in public places: Songs of black vocal rhythm and blues or doo-wop', *Journal of Aesthetics and Art Criticism*, 71(1): 101–110

Goodwin, Daisy. 2012. 'Venus de Hollywood: Marilyn Monroe at Salvatore Ferragamo Museum', *Newsweek*, 27 August <https://www.newsweek.com/venus-de-hollywood-marilyn-monroe-salvatore-ferraga mo-museum-64533>

Gracyk, Theodore. 1996. *Rhythm and Noise* (Durham and London: Duke University Press)

——. 2001. *I wanna be me: Rock music and politics of identity* (Philadelphia: Temple University Press)

——. 2007. 'Allusions and intentions in popular art', in *Philosophy and the Interpretation of Pop Culture*, ed. by William Irwin and Jorge J. E. Gracia (Lanham, Maryland: Rowman & Littlefield), pp. 65–87

——. 2012/3. 'Covers and communicative intentions', *The Journal of Music and Meaning*, 11: 22–46 <https://jmm11.musicandmeaning.net/>

——. 2021. 'The philosophy of jazz, popular music and art', interview by Richard Marshall <https://www.3-16am.co.uk/articles/the-ph ilosophy-of-jazz-popular-music-and-art?c=end-times-serie s>

Greene, Doyle. 2014. *The Rock Cover Song: Culture, History, Politics* (Jefferson,

North Carolina: McFarland & Company)

Griffths, Dai. 2002. 'Cover versions and the sound of identity in motion', in *Popular Music Studies*, ed. by David Hesmondhalgh and Keith Negus (London: Arnold), pp. 52–64

Haas, Riley. 2020. 'Michelle Kash takes "Personal Jesus" from sinister to sexy', *Cover Me* blog, 23 January <https://www.covermesongs.com /2020/01/michelle-kash-personal-jesus-cover.html>

Hamm, Charles. 1994. 'Genre, performance and ideology in the early songs of Irving Berlin', *Popular Music*, 13(2): 143–150

Hirsch, Lily E. 2020. *Weird Al: Seriously* (Lanham, Maryland: Rowman & Littlefield)

Hodge, Will. 2021. 'Social Distortion's Mike Ness on his 10 best country-punk covers', *Rolling Stone*, 23 March <https://www.rollingstone .com/music/music-country-lists/social-distortions-mike-n ess-on-his-10-best-country-punk-covers-110874/>

Horn, David. 2000. 'Some thoughts on the work in popular music', in *The Musical Work: Reality or Invention?*, ed. by Michael Talbot (Liverpool University Press), pp. 14–34

Hull, David L. 1976. 'Are species really individuals?', *Systematic Zoology*, 25(2): 174–191

——. 1978. 'A matter of individuality', *Philosophy of Science*, 45(3): 335–360

Inglis, Ian. 2005. 'Embassy Records: Covering the market, marketing the cover', *Popular Music and Society*, 28(2): 163–170

Irwin, William. 2001. 'What is an allusion?', *Journal of Aesthetics and Art Criticism*, 59(3): 287–297

——, ed. 2013. *Black Sabbath and Philosophy: Mastering Reality* (Chichester, West Sussex: John Wiley & Sons)

JJJ (Triple J). 2021a. 'Behind JK-47's "Changes" Like a Version (interview)', *YouTube*, 11 February <https://www.youtube.com/watch?v=VtQ8k rwI7x0>

——. 2021b. 'Behind Something for Kate's Taylor Swift Like a Version (interview)', *YouTube*, 29 July <https://www.youtube.com/watch?v=2 E8Kfdfbp1M>

Jones, Gaynor, and Jay Rahn. 1977. 'Definitions of popular music: Recycled', *The Journal of Aesthetic Education*, 11(4): 79–92

Kania, Andrew. 2006. 'Making tracks: The ontology of rock music', *Journal of Aesthetics and Art Criticism*, 64(4): 401–414

——. 2020. *The Philosophy of Western Music: A Contemporary Introduction* (New York and London: Routledge)

Kieran, Matthew. 2008. 'Why ideal critics are not ideal: Aesthetic character, motivation and value', *The British Journal of Aesthetics*, 48(3): 278–294

Kirschner, Tony. 1998. 'Studying rock: Towards a materialist ethnography', in Swiss et al., pp. 247–268

Kripke, Saul A. 1972 *Naming and Necessity* (Cambridge, Massachusetts: Harvard University Press)

Kurp, Josh. 2015 'Gwar seamlessly covers Cyndi Lauper's "She Bop" and the Ramones' "Blitzkrieg Bop"', *Uproxx*, 27 October <https://uproxx.com/music/gwar-she-bop-cover-video/>

Lapriore, Elaine Beebe. 2001. '"Teen Spirit" at 10: An unshakable scent', *The Washington Post*, 2 September <https://www.washingtonpost.com/archive/lifestyle/style/2001/09/02/teen-spirit-at-10-an-unshakable-scent/e76f5ad4-17ad-4c34-9479-0bc5194ad402/>

Leddington, Jason P. 2021. 'Sonic pictures', *Journal of Aesthetics and Art Criticism*, 79(3): 354–365

Lenig, Stuart. 2010. 'David Bowie's Pin-Ups: Past as prelude', in Plasketes, pp. 127–136

Leszczak, Bob. 2014. *Who Did It First? Great Rock and Roll Cover Songs and Their Original Artists* (Plymouth, UK: Rowman & Littlefield)

Light, Alan. 2012. *The Holy or the Broken* (New York: Atria Books)

Londergan, Tim. 2018. 'Oh boy! Sonny West; the Crickets; Brian Setzer', *Tim's Cover Story*, 5 November <https://timscoverstory.wordpress.com/2018/11/05/oh-boy-sonny-west-the-crickets-brian-setzer/>

Lucifer. 2018. 'The last heartbreak', season 3, episode 18

Mag Uidhir, Christy. 2007. 'Recordings as performances', *British Journal of Aesthetics*, 47(3): 298–314

Magnus, Cristyn, P.D. Magnus, and Christy Mag Uidhir. 2013. 'Judging covers', *Journal of Aesthetics and Art Criticism*, 71(4): 361–370

Magnus, Cristyn, P.D. Magnus, Christy Mag Uidhir, and Ron McClamrock. 2022. 'Appreciating covers', *Nordic Journal of Aesthetics*

Magnus, P.D. 2008. 'Mag Uidhir on performance', *The British Journal of Aesthetics*, 48(3): 338–345

——. 2012. *Scientific Enquiry and Natural Kinds: From Planets to Mallards* (Basingstoke, Hampshire: Palgrave MacMillan)

——. 2013. 'Historical individuals like Anas platyrhynchos and "Classical gas"', in *Art & Abstract Objects*, ed. by Christy Mag Uidhir (Oxford University Press) pp. 108–124

Malawey, Victoria. 2014. '"Find out what it means to me": Aretha Franklin's gendered re-authoring of Otis Redding's "Respect"', *Popular Music*, 33(2): 185–207

McCormick, Neil. 2007. 'Paul Anka: One song the Sex Pistols won't be singing', *The Daily Telegraph*, 8 November, p. 29

McDonald, Tom. 2021. 'Good, Better, Best: The Smiths' "How Soon Is Now?"', *Cover Me*, 15 March <https://www.covermesongs.com/2021/03/good-better-best-smiths-how-soon-is-now.html>

McLean, Don. 2004. 'On the incorrect use of the term "cover"', 26 August, web page recovered via the Internet Archive <https://www.don-mclean.com/news/news_item.asp?NewsID=77> archived at <https://web.archive.org/web/20060620100451/http://www.don-mclean.com/news/news_item.asp?NewsID=77>

McLeish, Claire E. A. 2020. *"All Samples Cleared!": The Legacy of Grand Upright v. Warner in Hip-hop, 1988–1993*, PhD thesis, Schulich School of Music, McGill University, Department of Music Research, April

Metcalf, Greg. 2010. 'The same yet different/different yet the same: Bob Dylan under the cover of covers', in Plasketes, pp. 177–187

Mill, John Stuart. 1873 (2003). *Autobiography* (Project Gutenberg) <https://www.gutenberg.org/ebooks/10378>

Millikan, Ruth Garrett. 1984. *Language, Thought, and Other Biological Categories: New Foundations for Realism* (Cambridge, Massachusetts: The MIT Press)

Montgomery, James. 2007. 'Wu-Tang Clan's "first-ever cleared Beatles sample" claim is incorrect', MTV News, 3 October <https://www.mtv.com/news/1571114/wu-tang-clans-first-ever-cleared-beatles-sample-claim-is-incorrect/>

Mosser, Kurt. 2008. '"Cover songs": Ambiguity, multivalence, polysemy', *Popular Musicology Online*, 2 <http://www.popular-musicology-online.com/issues/02/mosser.html>

Nado, Jennifer. 2021. 'Conceptual engineering, truth, and efficacy', *Synthese*, 198(7): 1507–1527

NRPB, National Recording Preservation Board. 2018. 'Registry titles with descriptions and expanded essays', Library of Congress <https://ww

w.loc.gov/programs/national-recording-preservation-board/r
ecording-registry/descriptions-and-essays/>

Neely, Adam. 2021. 'Did Olivia Rodrigo steal from Paramore?', *YouTube*, 30
August <https://www.youtube.com/watch?v=qX7a2p5_JsM>

NY Times. 1976. 'George Harrison guilty of plagiarizing, subconsciously, a
'62 tune for a '70 hit', 8 September, p. 42

Noonan, Tom. 1965. 'Cover war raging—with share-the-wealth twist', *Billboard*, 27 February, pp. 1, 10

Nussbaum, Charles O. 2007. *The Musical Representation: Meaning, Ontology,
and Emotion* (Cambridge, Massachusetts: The MIT Press)

——. 2021. 'Ontology', in *The Oxford Handbook of Western Music and Philosophy*, ed. by Tomás McAuley, Nanette Nielsen, and Jerrold Levinson
(Oxford University Press) pp. 325–344

Padgett, Ray. 2017. *Cover Me: The Stories Behind the Greatest Cover Songs of
All Time* (New York: Sterling)

——. 2020. *I'm Your Fan: The Songs of Leonard Cohen* (Sterling, Bloomsbury
Academic)

Plasketes, George, ed. 2010. *Play it again: Cover songs in popular music* (Ash-
gate)

Plaugic, Lizzie. 2015. 'Sounds like a hit: the numbers game behind Spotify
cover songs', *The Verge*, 8 September <https://www.theverge.com/2
015/9/8/9260675/spotify-cover-songs-taylor-swift-adele>

Polite, Brandon. forthcoming. 'Taylor Swift, *Fearless (Taylor's Version)*', *Bloomsbury Contemporary Aesthetics*

Popdose Staff. 2011. 'The Popdose 100: The greatest cover songs of all time',
31 August <https://popdose.com/the-popdose-100-the-greate
st-cover-songs-of-all-time/>

Prato, Paolo. 2007. 'Selling Italy by the sound: Cross-cultural interchanges
through cover records', *Popular Music*, 26(3): 441–462

Prinz, Jesse. 2014. 'The aesthetics of punk rock', *Philosophy Compass*, 9(9):
583–593

Proximo, Dr. 2017. 'Songs you didn't know were covers', 27 August <http:
//www.geeksandbeats.com/2017/08/songs-didnt-know-covers/>

Pulitzer.org. 2008. 'Bob Dylan', The Pulitzer Prizes <https://www.pulitz
er.org/winners/bob-dylan>

Putman, Daniel A. 1982. 'Natural kinds and human artifacts', *Mind*, 91(363):
418–419

Putnam, Hilary. 1975. 'The meaning of "meaning"', in *Minnesota Studies in Philosophy of Science*, volume VII, ed. by Keith Gunderson (Minneapolis: University of Minnesota Press), pp. 131–193

Richard, Steven. 2018. 'Chris Cornell mashed up U2 and Metallica's "One"', *Unruly Stowaway*, 16 September <https://www.unrulystowaway.com/chris-cornell-u2-metallica-one/>

Rings, Michael. 2013. 'Doing it their way: Rock covers, genre, and appreciation', *Journal of Aesthetics and Art Criticism*, 71(1): 55–63

Rivers, Tony. 2007. 'Top of the Pops', personal website recovered via the Internet Archive <http://www.tonyrivers.com/totp.html> archived at <https://web.archive.org/web/20070706141310/http://www.tonyrivers.com/totp.html>

Rohrbaugh, Guy. 2003. 'Artworks as historical individuals', *European Journal of Philosophy*, 11(2): 177–205

Rolling Stone. 2011. 'Rolling Stone readers pick the top 10 greatest cover songs', 2 March <https://www.rollingstone.com/music/music-lists/rolling-stone-readers-pick-the-top-10-greatest-cover-songs-12792/>

——. 2012. 'Def Leppard re-recording "forgeries" of old hits', 13 July <https://www.rollingstone.com/music/music-news/def-leppard-re-recording-forgeries-of-old-hits-247079/>

——. 2021. '500 greatest songs of all time', 15 September <https://www.rollingstone.com/music/music-lists/best-songs-of-all-time-1224767/>

Ross, Stephanie. 1981. 'Art and allusion', *Journal of Aesthetics and Art Criticism*, 40(1): 59–70

Siese, April. 2016. 'The mystery man who turns every song into Smash Mouth's "All Star"', *Daily Dot*, 21 October <https://www.dailydot.com/unclick/smashmouth-all-star-mashups-jon-sudano/>

Simon, Bill. 1949. 'Small labels' ingenuity and skill pay off', *Billboard*, 3 December, pp. 3, 13, 18

Simpson, Dave. 2019. 'How we made I think we're alone now: Tommy James and Tiffany on their shared hit', *The Guardian*, 30 July <https://www.theguardian.com/music/2019/jul/30/tommy-james-shondells-tiffany-how-we-made-i-think-were-alone-now>

Snow, Mat. 1988. 'Leonard Cohen: Cohen's way', *The Guardian*, February, archived at Rock's Backpages

Solis, Gabriel. 2010. 'I did it my way: Rock and the logic of covers', *Popular Music and Society*, 33(3): 297–318

Sparks, Hannah. 2021. 'Olivia Rodrigo adds Paramore writing credit to "Good 4 U"', *New York Post*, 25 August <https://nypost.com/2021/08/25/olivia-rodrigo-adds-paramore-writing-credit-to-good-4-u/>

St. Pierre, Roger. 1975. 'Worries of the Warwick sisters', *New Musical Express*, 12 July

Swiss, Thomas, John Sloop, and Andrew Herman, ed. 1998. *Mapping the beat: Popular music and contemporary theory* (Blackwell)

Telegraph. 2004. 'They did it their way', the Telegraph's music critics select the 50 best cover versions ever recorded, *The Telegraph*, 20 November,

Thomerson, John P. 2017. *Parody as a Borrowing Practice in American Music, 1965–2015*, PhD thesis, University of Cincinnati, Division of Composition, Musicology, and Theory

Travers, Mary. 1975. 'Mary Travers and friend', aired on KNX-FM (Los Angeles) 20 April; printed in *Dylan on Dylan*, ed. by Jeff Burger (Chicago Review Press) 2018

Treble Staff. 2018. 'The top 100 cover songs', *Treble*, 25 July <https://www.treblezine.com/top-100-best-cover-songs/>

Velez, Denise Oliver. 2021. 'On original Black music, successful white cover songs, and a culture of covetousness and cruelty', *Daily Kos*, 27 June <https://www.dailykos.com/stories/2021/6/27/2036225/-On-original-Black-music-successful-white-cover-songs-and-a-culture-of-covetousness-and-cruelty>

Wald, Elijah. 2009. *How the Beatles destroyed Rock 'n' Roll: An alternative history of American popular music* (Oxford University Press)

Weinstein, Deena. 1998. 'The history of rock's pasts through covers', in Swiss et al., pp. 137–151.

——. 'Appreciating cover songs: Stereophony', in Plasketes, pp. 243–251

What The France. 2020. 'Once upon a song: The story of "Comme d'habitude", which became the international hit "My way"', *What the France*, 11 June <https://whatthefrance.org/my-way-the-legendary-song-celebrates-its-50-year-birthday-bertrand-dicale-meets-jacques-revaux-for-what-the-france/>

Willman, Chris. 2021. 'Taylor Swift's "Fearless (Taylor's version)" debuts huge: What it means for replicating oldies, weaponizing fans', *Variety*,

20 April <https://variety.com/2021/music/news/taylor-swift-fearless-lessons-1234955475/>

Woolworths. 2017. 'Embassy records: The cover story', *The Woolworths Museum* <http://www.woolworthsmuseum.co.uk/1950s-embassyrecords.htm>

Zak, Albin J., III. 2001. *The Poetics of Rock: Cutting Tracks, Making Records* (Berkeley: University of California Press)

——. 2010. *I Don't Sound Like Nobody: Remaking Music in 1950s America* (Ann Arbor: University of Michigan Press)

Zollo, Paul. 2012. *Conversations With Tom Petty* (Omnibus Press)

Index

About the Team

Alessandra Tosi was the managing editor for this book.

Rosalyn Sword performed the copy-editing and proofreading.

Anna Gatti designed the cover. The cover was produced in InDesign using the Fontin font.

P.D. Magnus typeset the book in LaTeX and produced the paperback, hardback, PDF and EPUB editions. The text font is Tex Gyre Pagella; the heading font is Californian FB.

Luca Baffa produced the AZW3, HTML, and XML editions—the conversion is performed with open source software such as pandoc (https://pandoc.org/) created by John MacFarlane and other tools freely available on our GitHub page (https://github.com/OpenBookPublishers).

This book need not end here...

Share

All our books — including the one you have just read — are free to access online so that students, researchers and members of the public who can't afford a printed edition will have access to the same ideas. This title will be accessed online by hundreds of readers each month across the globe: why not share the link so that someone you know is one of them?

This book and additional content is available at:

https://doi.org/10.11647/OBP.0293

Donate

Open Book Publishers is an award-winning, scholar-led, not-for-profit press making knowledge freely available one book at a time. We don't charge authors to publish with us: instead, our work is supported by our library members and by donations from people who believe that research shouldn't be locked behind paywalls.

Why not join them in freeing knowledge by supporting us: https://www.openbookpublishers.com/section/104/1

Like Open Book Publishers

Follow @OpenBookPublish

Read more at the Open Book Publishers BLOG

You may also be interested in:

Auld Lang Syne
A Song and its Culture
M. J. Grant

https://doi.org/10.11647/OBP.0231

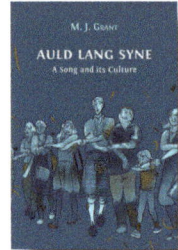

Acoustemologies in Contact
Sounding Subjects and Modes of Listening in Early Modernity
Emily Wilbourne and Suzanne G. Cusick (eds)

https://doi.org/10.11647/OBP.0226

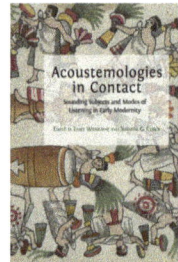

www.ingramcontent.com/pod-product-compliance
Lightning Source LLC
Chambersburg PA
CBHW050809270326
41926CB00026B/4652